Unexpected Poison

People go to work hoping for a reliable environment and trustworthy leadership who will provide what is needed for them to work well. This hope is true for any worker, especially those whose work entails exposure to physical or psychological hazards in the service of their community. High rates of 'burnout' and psychological distress are becoming increasingly common across all professions. This book explores the impact of harmful leadership practices and workplace cultures on staff while the organisation publicly projects an image of high ethical standing. It investigates those at the frontline of work whose expectations of organisational safety and fairness have not only been breached but betrayed.

This book describes 'moral injury', a new way to understand the impact of the 'betrayal of what is right by someone in legitimate authority'. This interdisciplinary title presents first-person accounts of those who have experienced betrayal in the workplace and it contains reflections of those accounts from numerous angles so that readers can understand how they too may be impacted by similar experiences. Utilising psychology, philosophy, theology, and sociology, readers are given an understanding of the latest research on moral injury so they may avoid the damaging impacts that broken workplace cultures and betrayal can have on their own lives and moral virtue. It provides signposts to strategies that move towards 'the good' by avoiding the harmful impacts of resentment and vengeance on their betrayers and, instead, it offers a holistic framework that repairs broken systems and instils hope, healing and a way to grow after the event.

Unexpected Poison: Betrayal and Moral Injury in the Workplace is a fascinating read for health and safety practitioners and any professional in a high-risk industry including aviation, medicine, military, construction, transport, and engineering.

Unexpected Poison

Betrayal and Moral Injury in the Workplace

Mark Layson

CRC Press
Taylor & Francis Group
Boca Raton London New York

CRC Press is an imprint of the
Taylor & Francis Group, an informa business

Designed cover image: Mark Layson

First edition published 2026
by CRC Press
2385 NW Executive Center Drive, Suite 320, Boca Raton FL 33431

and by CRC Press
4 Park Square, Milton Park, Abingdon, Oxon, OX14 4RN

CRC Press is an imprint of Taylor & Francis Group, LLC

ISBN: 9781032814094 (hbk)
ISBN: 9781032803067 (pbk)
ISBN: 9781003499725 (ebk)

DOI: 10.1201/9781003499725

Typeset in Times
by Newgen Publishing UK

Dedication

To my wife and family, without whom I would lack encouragement and the provocation to virtue.
For my shepherd, who leads to pleasant places and heals dented souls.

Contents

PART 3 A Better Narrative

Foreword

Some of you will be familiar with the genre: books that trace a drift into moral compromise, betrayal of values and resulting organisational harm with clinical precision (Mandis, 2013; Mayer, 2009; McLean & Elkind, 2004; Snook, 2000; Vaughan, 1996). Some of these books, more than others, speak to the human heart that beats beneath the insidious growth and infliction of such harm. What Mark Layson has written here does both. *Unexpected Poison* is a courageous and urgently necessary attempt to name what Bonhoeffer (1995) already knew: that not all wounds in the workplace are physical, and not all injuries can be healed with policies, mindfulness apps, or new slogans on a poster. As many workers know themselves from experience: some injuries go deeper and resist an easy patch-up. These are the kinds of injuries and betrayals that make the moral universe wobble. They unsettle our sense of what is right. They cheat on something we thought we could trust. They disturb the ethical architecture of our working life.

Moral injury is one of those difficult truths that organisations have long resisted naming, partly because naming it carries responsibility. To acknowledge moral injury is to admit that the organisation itself can betray and wound—not merely through isolated mistakes or ill-considered responses to failure, but by demanding that its people make every-day moral compromises, that they sideline their conscience, that everybody submit to managerial mechanics that look great on a spread sheet but that permit betrayal both to occur and go unaddressed. The great value of *Unexpected Poison* is that it refuses to soften that truth. Mark brings the reader face to face with it, with honesty and with empathy, but without pity.

Of course I have my own proclivities when I read a book like this, but what strikes me most is that *Unexpected Poison* is ultimately a book about compassion (or its absence). What I mean is not compassion as sentiment or softness, but as an acceptance of responsibility and an expression of moral seriousness. Compassion, properly understood, demands that we respond to suffering with truth, responsibility, and agency. The kind of compassion Mark writes about is not the anaemic corporate wellbeing version that puts the burden of healing back onto the harmed. It is compassion as moral work: the work of acknowledging that injuries have been inflicted, the work of attending to the injured, of repairing what is broken, and restoring the human and social dignity that organisational betrayal and harm have taken away.

In my own writing on forgiveness and organisational response to error, I have argued that forgiveness is never an escape hatch from responsibility. It is not a psychological trick we offer the injured to help them 'move on,' nor a managerial shortcut to closure. Forgiveness, where it is possible, grows only in the soil of truth-telling, accountability, and change. The same is true of healing from moral injury. This book makes that point with clarity: healing is never cheap. Trust cannot be rebuilt by following a process more closely or by listening to a couple of instructions of how to behave toward one another. Similarly, closure after the infliction of moral injury cannot be rushed into existence—genuine closure runs on a timeline unfurled by *kairos*, not *chronos*. A harmed worker does not recover because they have been told to be resilient and more trusting. No, they (and their colleagues) can begin to recover when the organisation becomes worthy of their trust and their resilience again.

Mark's interdisciplinary approach—an interweaving of psychology, ethics, theology, anthropology, and the lived experiences of workers—presents us with a vocabulary that, I know, many have been searching for. This vocabulary helps us see the hidden mechanisms by which workplaces erode integrity. It illuminates the quiet virtues that help people endure harm with a kind of grace which they should never have been asked to muster. But more importantly, it offers a map for organisational leaders who genuinely want to do better. Moral injury is not inevitable. It is the predictable outcome of clashing incentive structures and organisational habits. These *can* be changed—with courage, humility, and moral imagination.

Mark's book is not simply a scholarly or empirical contribution to the literature. It is an invitation. An invitation for leaders to treat organisational life as a moral practice. An invitation for workers to recognise their experiences in these pages and to know they are not alone. And an invitation for all of us to build workplaces where compassion is not a program, but a posture (or poise, as Mark would probably say); where accountability is not punitive, but restorative; and where we have the courage to tell the truth about the harm we cause—not to shame or blame, but to heal.

With *Unexpected Poison,* Mark has done something rare. He has pulled betrayal and moral injury out from the shadows and placed them into the everyday vocabulary of working life. This is where they urgently belong. And he has done so with the depth, humility, and moral clarity of someone who has done the hard yards on the frontline, who has lived alongside suffering and has increasingly paid attention to it. That alone makes this book worth reading.

Yet the greatest gift is probably this: *Unexpected Poison* helps us imagine a better way. Mark shows us a more compassionate way. A more responsible way. A more human way. And in a time when workplaces are too often defined by quarterly financial obsessions, haste, indifference, or denial, that is beautiful and acutely necessary.

Sidney Dekker, MA, MSc, PhD
Professor, Crisis Chaplain
Griffith University, Brisbane, Australia

REFERENCES

Bonhoeffer, D. (1995). *Ethics*. Simon & Schuster.

Mandis, S. G. (2013). *What happened to Goldman Sachs: An insider's story of organizational drift and its unintended consequences.* Harvard Business Review Press.

Mayer, J. (2009). *The dark side*. Knopf.

McLean, B., & Elkind, P. (2004). *The smartest guys in the room: The amazing rise and scandalous fall of Enron*. Portfolio. http://www.loc.gov/catdir/enhancements/fy0720/2004541366-b.html

http://www.loc.gov/catdir/enhancements/fy0720/2004541366-d.html

Snook, S. A. (2000). *Friendly fire: The accidental shootdown of US Black Hawks over Northern Iraq*. Princeton University Press.

Vaughan, D. (1996). *The Challenger launch decision: Risky technology, culture, and deviance at NASA*. University of Chicago Press. http://www.loc.gov/catdir/description/uchi051/95039858.html
http://www.loc.gov/catdir/toc/uchi051/95039858.html
http://www.h-net.org/review/hrev-a0a3o8-aa

Preface

THE WITCHES' BREW

CRUSHING IT

Mike is the kind of person you want alongside you in a tough time. He is a gentleman in the true sense of the word. Hardworking, unassuming of others, not seeking wealth or fame – someone striving to make a positive difference in the world. His drive to bring positive change was part of his rural family's DNA; it was central to his family's deepest beliefs about themselves and their small-town community. They all entered serving professions and prolifically volunteered in their hometown. Mike began volunteering before he entered his teenage years, and in adulthood, he pursued a career path centred on service to his community. As he aged, Mike looked back with a sense of fulfilment and even a little pride.

Mike came to me after a soul-denting experience working for a not-for-profit provider in the disability sector, supporting clients with significant and diverse needs. The organisation painted itself as compassionate, one that cared for the vulnerable wherever they might be found. Mike had a supportive line manager and started in a large work site, where another manager oversaw operations. Initially, Mike found the work rewarding because he was able to make a positive difference in the lives of those he encountered.

Slowly, though, Mike noticed something wasn't quite right – there was a culture of fear in the other employees. Noting Mike's concern for staff and clients, operational staff began to approach him confidentially. Over a dozen staff members, mostly poorly paid, ushered Mike into the few unsurveilled corners of the worksite for hushed conversations.

In those hidden corners, staff whispered about the aggressive and punitive style of the site manager. Some reported threats and aggression; others, pressure to sign unreasonable workplace agreements that would remove their entitlements. Mike discovered that these threats were the business-as-usual management style at this site, but strongly at odds with how the organisation painted itself.

As we talked, Mike flicked through the notes he made at the time, recalling that in his hopes of calming the situation, he'd made a plan of action in consultation with his own line manager. Mike met his line manager at a local café to check he wasn't overreading the situation or seen to be acting as a lone ranger. The management style at the site, as it turned out, was a well-known secret that had never been addressed. With a wry smile, the line manager agreed to support Mike in what would be a complex process. Within a week, the situation at the site worsened for Mike. Other junior managers started avoiding contact with Mike, and he was silently removed from the necessary workplace computing systems.

Mike was glad he had discussed his plan beforehand and reached out for help from his line manager. However, instead of the support he had been promised, Mike's line manager started ghosting him, ignoring emails and requests for meetings. When the manager eventually answered, it was in emails filled with corporate speak and defensiveness. Finally, a meeting time and place were agreed upon between Mike and his line manager. Mike was grateful to finally receive the promised support.

Mike was ushered into a deserted workstation. 'They will crush us!' Mike's manager spoke in a faltering voice that was indicative of genuine fear. 'You need to drop this issue now. If you don't, they will destroy your reputation. You will be made out as the problem… my career and our program will be crushed'. Before his eyes, Mike saw any support evaporate, replaced instead by the harsh reality of a poisoned chalice.

The organisational system was failing Mike, but simultaneously working as designed. Organisations may assert that they are open and inclusive; however, their systems most frequently work to protect the organisation, and not its employees. This organisational self-protective inclination is the central feature and central failure of most workplaces. The glossy value statements often work one way only. This self-protective system ensured Mike would be removed from duties within half a day and lose his job within two weeks, while the site manager received an award for high compliance ratings.

Like many who have been failed or betrayed by a self-protective bureaucracy, Mike's first reaction was visceral – his heart dropped, his guts churned, his mind raced, and his hands shook. The poisonous response of his organisation was utterly unexpected. Mike began to feel a growing anger at the injustice being done to staff, which now included him. Not many people expect virtuous organisations to behave so poorly. When they do, the psychological and emotional sensations experienced are like those of people who've been stabbed in the back by a trusted ally. Mike was experiencing an act of organisational betrayal – and it hurt.

You may have picked up this book because something like Mike's experience has happened to you or someone you know. You recognise Mike's story, but you may not have had the right words to explain what happened, why it happened, or the pain it caused you. This book will give you new language to understand the machinations of workplace betrayal and the resultant sting of moral injury. It will allow you to understand your experiences afresh so you can retell your story with an ending that produces hope and meaning.

Workplace betrayals like Mike's are common, although seldom discussed with any real depth, because the current language used to describe workplace distress and psychological injury does not yet accommodate betrayal and moral injury. The current paradigm used for workplace mental and emotional stress is failing and needs serious reconsideration, if not replacement. It requires change because the statistics on workplace mental health tell a tale of increasing distress.

The rise of workplace psychological injuries is directly proportional to the amount of money spent to prevent and treat them. That is, the more money is spent to address workplace distress, the worse the situation becomes. Work-related anxiety, depression, burnout, and post-traumatic stress disorder (PTSD) are increasing at similar rates to the introduction of resilience training, mental first aid, and Employee Assistance Programmes. Australian organisations have gradually increased annual spending on wellness initiatives to AUD 300 million. Still, over this decade, there has been a 97% increase in workplace mental health claims (Safe Work Australia, 2024).

It is not unreasonable to conclude that current workplace mental health initiatives are generally ineffective.

Many factors contribute to this costly irony; crucial, however, is that many organisations (and researchers) do not comprehend what is causing poor workplace mental health. In a recent review of the mental health programmes in New South Wales Police (NSWP), the Auditor General noted the upward trend in money spent to solve a worsening problem. The report concluded that most of their programmes were ineffective because they had been 'established in the absence of analysis of the factors that lead to [...] psychological injuries' (Auditor-General of NSW, 2025).

Like most organisations, the NSWP addresses workplace mental health from a 'wellbeing' perspective, targeting individual fragility, rather than addressing the issue at an organisational systems level. Mike had access to workplace wellbeing programmes, like cupcake morning teas, exercise lessons, and resilience workshops. Yet, little was done to address systemic leadership abuse at the root of staff distress.

The individual wellbeing mindset is promoted and pursued with religious zeal and capitalistic drive. Globally, corporations spend more on workplace wellbeing programmes than the entire annual healthcare budget of Australia each year. Courage and humility are required to rethink the causes and solutions of increasing levels of distress in workers – change is long overdue.

Rethinking Workplace Wellbeing from the Ground Up

In 1858, the Great Stink of London was in full swing. The River Thames was a toxic stew of pollutants that gurgled and belched its way through the heart of London. The air above was so pungent that sessions in Parliament were paused. Worse than the stink was the toll from the associated outbreak of cholera, commonly considered to be an airborne disease carried by an evil miasma. The cause of cholera was simplistically assumed to be the all-pervading stink that disturbed the city and its residents.

With the unfounded miasma theory in full swing, pseudo-medical treatments were routinely given to those struck down with the disease. In the initial stages, treatments commenced with providing chamomile tea, and as symptoms worsened, tobacco smoke enemas were administered. In the desperate final stages, people were inflicted with cloth and beeswax anal plugs to stop diarrhoea! Other than offering these futile treatments, many felt helpless to stop its impacts, but people should not have felt powerless.

Nearly ten years prior, a London physician named John Snow challenged the status quo in a pamphlet, *On the Mode of Communication of Cholera* (Snow, 1849). The publication challenged the miasma theory, stating that cholera was, in fact, a water-borne disease. At the time, the germ theory was unknown to the world, and people couldn't comprehend what Snow was saying, so they largely ignored him. The poison was not where people assumed it to be. However, some other trailblazers and non-conformists were coming to believe that cholera was water-borne.

These troublemakers included an Anglican cleric, Rev Sydney Smith, who wrote in 1834: 'He who drinks a tumbler of London water has literally in his stomach more animated beings than there are men, women, and children on the face of the globe'. More important than Smith's quip were the efforts of engineer Joseph Bazalgette, who had a plan to prevent contact with infected water. One year after the Great Stink, he oversaw the construction of the London sewer system – an engineering marvel leading to the eradication of large-scale cholera outbreaks in London, and a legacy of health prevention around the globe. Intriguingly, the solution to a medical epidemic was an engineering solution.

Three relevant conclusions can be drawn. First, once ideas, even bad ones, become public, they are tough to shift. The miasma theory needed to be replaced by correct descriptions of cholera's cause. However, no one expected poison in the water, so people persisted for years with common thought rather than considering something less obvious. Many lives would have been spared if people like Snow, Smith, and Bazalgette had been given a proper hearing earlier.

Addressing the spiralling issue of workplace mental health through *miasmic* wellbeing interventions also requires the courage to *Think Again* (Grant, 2023). Organisational Psychologist, Adam Grant, encourages us to unleash the power of unknowing so that we might unshackle ourselves from older, unhelpful, and destructive ideas. When organisational leaders embrace with humility and courage the notion that they might be wrong about worker wellbeing, we will move from ineffective individual programmes towards effective systems.

Second, the Great Stink shows that understanding the cause of a problem and describing it well helps you find an effective solution. The solutions for miasmas were harmless, strange, or sometimes worse than the disease itself. The correct solution for cholera was counterintuitive and odd compared to the accepted wisdom of the time. The re-thinkers looked foolish when they said, 'People are sick – let's do some engineering'.

Bazalgette and others developed a systemic solution that prevented the disease instead of trying to fix individual symptoms. Such systemic thinking and dialogue are necessary today. The right way to address workplace mental health issues, this book argues, is through changing harmful organisational cultures and the leaders who create, maintain, and thrive within poisonous and morally injurious cultures.

Third, cholera required a group of people with the ability to communicate and work collaboratively across disciplines to create systemic solutions. It sounds like the beginning of a joke, but when a doctor, cleric, and engineer aligned themselves on cholera, they were a mighty power to be reckoned with. It is increasingly recognised that the siloed solutions of uni-disciplinary experts seeking to address workplace mental health issues will fail. However, when an interdisciplinary team works together, new wonders can be created that will surpass the London sewer system.

'You Are the Problem!' Let's Rethink That

What lies at the heart of the wellbeing miasma is the belief that individual workers are almost solely to blame for their poor mental health. The following message is communicated: 'You are feeling stressed and overwhelmed at work, and YOU are the problem. If you were more resilient, did more yoga and had a better diet, there would be no problem'. It may not be communicated this bluntly (and I agree individuals should take responsibility for their own health); however, blame is communicated to forlorn staff from management and subject matter experts – like blaming fish for getting sick in polluted water.

To add insult to workplace psychological injury, organisations systemically white knight themselves, promoting their virtue (see Chapter 4) through their knowledge about wellbeing and their benevolence in funding programmes to fix self-broken workers. Organisations hand out online resilience training packages, wellbeing apps, cupcake morning teas (and ironically healthy cooking ideas), toll-free phone numbers, or radical options like ice baths and semi-naked snow hiking as solutions, much like chamomile tea, tobacco smoke enemas, and beeswax anal plugs of the 19th century.

Mike experienced this wellbeing miasma mindset amid his travails when he met with a Human Resources (HR) leader about the abusive manager. 'If we offered Tai-chi to the staff a few times a week, it might help' was the suggestion. This is a kind of 'wellbeing washing' of tokenistic support programmes painted as virtuous organisational support. Much of the wellbeing miasma is an industrial-scale gaslighting project (see Chapter 5) that blames workers rather than recognising harms caused by systems issues like workplace injustice and unsupportive leadership. Thankfully, the evidence is increasingly forcing a much-needed rethink.

In 2019, the Canadian researcher Dr Don McCreary completed an extensive worldwide review of wellbeing interventions in high-risk occupations. McCreary found that wellbeing interventions were overly focused on trying to fix individuals rather than change broken workplaces. McCreary also reported that no wellbeing programmes had solid evidence to support their claims of effectiveness in harm reduction (McCreary, 2019). McCreary was ahead of the curve, as in 2022, the World Health Organization (WHO) changed tack on workplace wellbeing programmes.

Driven by the fact that individual interventions had low to very low certainty of evidence for their effectiveness, only conditional support was given by WHO to individually focused workplace wellbeing interventions (World Health Organization, 2022). Instead, greater endorsement was given to systemic leadership and organisational interventions to help solve workplace mental health problems. A valid conclusion is that it is time to build systemic solutions to workplace distress that seek to fix broken workplaces, not just broken workers. Mike would have appreciated this changed emphasis.

Understanding Broken Workplaces

Like the convergence of ideas held by doctor, cleric, and engineer in 19th-century London, there is a growing group of thought leaders, such as McCreary, coming together from diverse fields to challenge the wellbeing miasma consensus. This book will cover much territory by detailing the latest research from a wide range of rethinkers – researchers, clinicians, carers, and even a priest or two. Together, this vanguard is helping craft a new language to understand and solve one of the most significant workforce dilemmas of our age – constructing a metaphorical sewer system to expunge harmful leadership and organisational practices which drive poor mental health in the workplace.

My own understanding of workplace health has taken a long time to develop. For over 30 years, like Mike, I have worked in supposedly kind, altruistic, or noble organisations that were not as kind and benevolent to their workers as they would have us believe. Indeed, the more noble the organisation, the more likely they are to leverage their public reputation to cover over organisational behaviour that belongs in a sewer. I knew something wasn't right in many of these workplaces, but I struggled to articulate the specific issues.

This book is my contribution to the construction of a new language founded on the concept of systemic workplace betrayal and the resultant harm caused through the infliction of moral injuries. I slowly came to understand this new language, but had a turning point when a friend (an Anglican priest) mentioned the concept of moral injury. This casual conversation eventually led me to leave my job and dedicate years to researching betrayal and moral injury, culminating in the completion of a PhD. This book reflects much of the content of my doctoral research.

I must pause to confess that I have not always been the perfect employer or employee. The research I conducted highlighted my own leadership failings, which have caused hurt to others. I am personally seeking to do better each day, and this book is not an attempt to stand and preach at others but is a dedication to the belief that failure today can lead to a better tomorrow, if only we would learn the lessons our errors can teach us.

A Brief Explanation of Moral Injury

The concept of moral injury may be new to many, so a brief explanation now (and a deeper dive in Chapter 2) will be helpful. Moral injury is an ancient concept, as old as betrayal itself; however, it has gained recognition since the mid-1990s. During the 1990s, the period of my twenties and the heyday of self-destructive themes in grunge music, pioneering psychiatrist Jonathan Shay addressed the lingering self-destructive damage occurring within the community of Vietnam Veterans. Shay became increasingly convinced that many veterans could recover from the horrors of conflict if their sense of right and wrong had not been breached. Notably, he highlighted that betrayal by leaders was the worst breach of right and wrong a soldier could experience. Shay stated that moral injuries were inflicted through the betrayal of what was right by those in legitimate authority in high-stakes situations (Shay, 2014).

Until now, moral injury and betrayal have mainly been researched in the context of 'high-stakes situations' involving traumatic events involving death or extreme/sexual violence, such as in war, policing, or life-or-death medical settings. The extreme nature of these contexts can make it hard to understand betrayal and moral injury occurring in the law firm, teacher's lounge, NGO, engineering company, church/temple/mosque or almost any place we work or derive meaning from.

This book, unlike many others on moral injury, is for the everyday person who may never venture into a battlefield. Civilian contexts involve high consequences because of the risk of losing one's livelihood and reputation, something neglected in the literature on moral injury. The mechanics of workplace betrayal are described in detail, and a new language is provided to describe these events.

Along with Mike's, are stories of others who have been betrayed in their workplace. These stories are anonymised because some who have told them are still in broken workplaces. The stories are then decoded using a language created by the latest research on workplace betrayal and moral injury. The studies surveyed come from a broad range of disciplines: lived experience, psychology, sociology, anthropology, ethics, philosophy, law, theology, and many more.

Using this interdisciplinary method allows a thick description of workplace betrayal and moral injury to be constructed. The solutions developed from these descriptions can address the full range of issues causing moral injuries. Wide-ranging expertise is needed because the reality of workplace betrayal is complex. Unlike academic specialisation, workplace betrayers rarely pick one theory or method when harming others. Instead, betrayers subconsciously choose from a buffet of harmful actions that help them achieve self-protection, which is experienced as betrayal by others. In our workplaces, betrayal and moral injury stem from a poisonous witches' brew of many concepts and theories.

A Witches' Brew

The ingredients that make for workplace moral injury are somewhat like the melting pot of the English language. English has a dollop of Greek, a serving of Latin, a splash of German, a pinch of French, along with many other ingredients – it is a witches' brew that has coalesced into a modern-day *lingua franca*. Likewise, our language of workplace betrayal and moral injury is not an isolated theory but a melting pot of other theories that coalesce into the negative experiences of workplace betrayal. To make sense of this, the book proceeds in three parts and ten chapters.

Part 1 will examine the foundational elements that make workers so vulnerable to betrayal, addressing the reader's own heart and soul. This initial focus on self is vital because, fundamentally, we must each understand ourselves. The aim of the book is not to blame leaders or organisations, but to heal us and the spheres we do and will influence. Part 1 unpacks personal meaning, identity, and our personal moral expectations about life and work. A deep dive into meaning is important because work is a two-edged sword to meaning. So often work creates and sustains our sense of meaning, purpose, and identity – when it is good, we thrive, but when it is broken, we languish. Part 1 teaches the most basic elements of moral injury and betrayal language, which function like the nouns, verbs, and adjectives of our new language.

Chapter 1 details how any work is a high-consequence context because it is a realm where humans build their identity and seek transcendent meaning. As a result, the workplace is intimately connected to deep belief structures. Chapter 2 explores three pairs of ideas that will help readers recognise the contours of workplace social structures. Firstly, *thick* and *thin* relationships are examined. Secondly, we note that overlapping frameworks of *ethics* and *morals* arbitrate these relationships. Finally, thick/thin relationships and morals/ethics are synthesised into a theory of *contractual* and *covenantal* workplace social structures. The expression of these three nuanced pairings sits at the foundation of a workplace where trust is built or is stagnant. It will be argued that contracts, although necessary legal structures, are insufficient to ensure workplace health. Chapter 3 thoroughly explores the concept of betrayal in history, the arts, and everyday life.

While Part 1 helps us understand ourselves, Part 2 helps us understand our workplaces. It further develops our language of moral injury by utilsing the nouns, verbs, and adjectives of our personal expectations and experiences of betrayal. It places them into a sentence structure, showing how the elements of our moral expectations are contorted in ways that create poisonous workplaces. We will explore how betrayal evolves into workplace betrayal, how organisations exploit contracts and covenants against their workers, and how deeply held moral expectations about our workplaces are manipulated, distorted, and compromised.

We will enrich our understanding of moral injury as a theory by discussing other concepts that contribute to a deep understanding of workplace betrayal. Some of the ingredients we will consider are psychological contract theory, psychological safety, psychosocial risks/hazards, institutional cowardice, organisational betrayal, the ancient art of paltering, the tactic of Deny-Attack-Reverse Victim-Offender (DARVO), gaslighting, and more. While this combination of ideas may seem overwhelming, we will work through it until we understand how they are deployed in isolation or combination to inflict moral injuries.

Chapter 4 takes a deep dive into the reputation and reality of the organisations we work for. Organisations mostly work to curate a public reputation of kindness and generosity, which is threatened by allegations of harmful internal cultures and practices. We will apply our learnings from Part 1 to recognise the halo effect in organisations and the organisational hubris that can form and lead to moral injuries.

Chapter 5 operationalises the concept of betrayal in an organisational context. This chapter is central to understanding the primary tactics used to manifest workplace betrayal. It will examine in-depth organisational self-protection instincts that manifest as institutional cowardice and the related concept of DARVO (Harsey & Freyd, 2020)

Chapter 6 examines the difference between truth, facts, lies, and workplace communications. Here, we will understand that truth and lies are not categorically defined, but instead exist on a spectrum. It is this ambiguity that can be leveraged by organisational representatives to deny malfeasance and cast blame elsewhere. Organisational representatives rarely tell outright lies. Instead, they make the most of the grey area between truth and lies to undertake the practice of paltering – the tool of choice when gaslighting staff and spinning organisational benevolence.

Chapter 7 builds on the content of preceding chapters to detail the scapegoating process. Scapegoating is the process of hanging organisational sin around the necks of 'bad apples' who have somehow infiltrated a good organisation. Scapegoats are often in the wrong place at the wrong time, bearing the blame for many organisational sins.

Finally, Part 3 takes a positive turn, using the elements of our new language for workplace betrayal (Part 1) and the sentence structure described to change the ending of the workplace betrayal narrative (Part 2). You are not the problem, and you don't have to be the helpless victim of betrayal. Those who have not learnt this language often hand over their personal moral agency to their betrayer in a spiral of anger and vengeance. However, we each have the power to rewrite our script to prevent the story of betrayal from coming to a destructive conclusion. Part 3 outlines mindset shifts and practical strategies that every reader can implement to defend their virtue and fix broken workplaces.

Firstly, it will help individual workers take responsibility for limiting their self-destructive actions, which they may justify due to past betrayals and injuries. An under-emphasised element of moral injury research is that those who feel betrayed often start to destroy their own character – by compromising their own belief structures, breaking laws, or even betraying others. Morally injured people break workplace rules, stone-wall colleagues, hurt their loved ones, and damage their own heart, soul, and mind. This is a process described by Jonathan Shay as the 'undoing of character'. A new language to describe the pain of moral injuries can help develop mechanisms to maintain our own moral virtue.

Secondly, Part 3 will guide workplace leaders, C-suite directors, HR departments and wellbeing experts who are willing to rethink. Contemplative leaders might use this new language to transform their organisations into ones that have constructive cultures and adaptive, even gracious leaders. More so, I hope it will help leaders to examine themselves and return from any moral degradation to the altruistic and kind motivations they had when they began.

The narrative of helping workers and mending workplaces unfolds across three chapters. Chapter 8 explores how moments of crisis often break the secrecy and spin upon which organisational betrayal relies, examining what former soldier, police officer, and educator Haydn McComas calls *Mea Culpa* moments, with the hope that poor workplace cultures might change by choice rather than through crisis.

Chapter 9 utilises our new language to help craft better personal narratives that feature virtue and integrity. Here we describe various practices to build virtue that are a simplified version of philosopher Mark Johnson's *Moral Imagination* and theologian Miroslav Volf's *Enlarged Thinking*. Both are designed to create personal awareness of the basic foundational elements described in Part 1, but ultimately unleash the power of forgiveness in constructing profound character, solid relationships, and safer workplaces. These practices will encourage personal and organisational virtues previously known as temperance, fortitude, faith, and justice, and help us turn away from self-destructive narratives of vengeance and blame towards ongoing and future-looking accountability for our actions.

Finally, Chapter 10 crafts a framework for better organisational narratives that mend broken workplaces. The framework prioritises organisational courage, psychosocial risk mitigation, shepherding leadership, Just Cultures, and, ultimately, cultures that value and enact forgiveness and reconciliation. Taking this journey to create positive cultures and leadership styles will not only prevent workplace betrayal but also

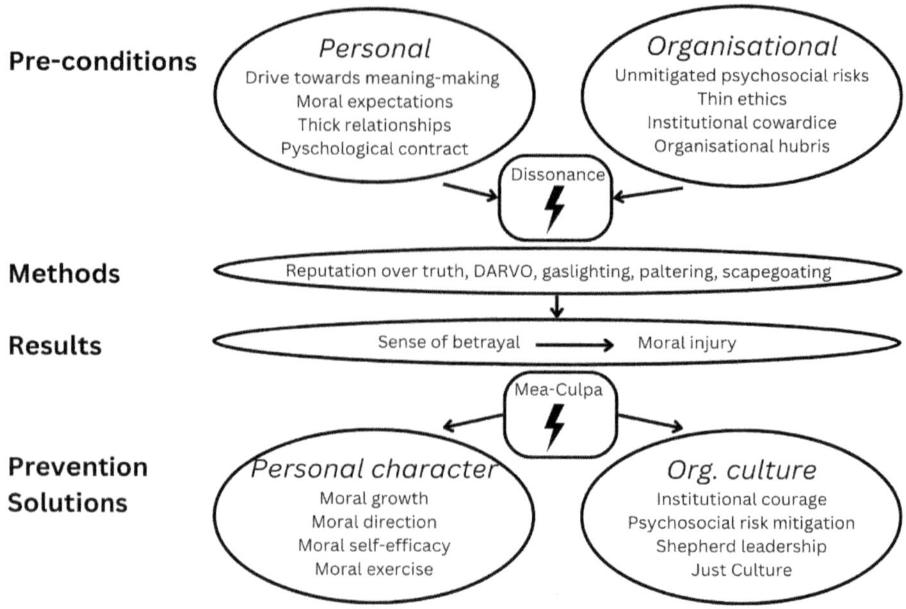

FIGURE 0.1 Witches' brew of concepts.

help reduce the costly impact of moral injuries that manifest as poor staff retention, low staff morale, and poor mental health of employees.

This witches' brew of betrayal and moral injury has many ingredients (see Figure 0.1). This diagram will be helpful to return to as we move from one element to another.

A Word of Cautious Optimism

Stories like Mike's are too familiar, and honestly, a little depressing. I am cautious because many people have expressed interest in sharing their anecdotes for this book. Conversely, after speaking to leaders of organisations all over the globe, I can see many squirming for ways to avoid these topics. I once presented significant data about betrayal and moral injury to a Chief Executive who relies heavily on wellbeing whitewash in the organisation he leads (with very poor wellbeing results). Instinctively, he replied, 'People will always look to find other things to blame'. It was an expression of organisational hubris (Chapter 4) that shut down rethinking before it started. This shutdown is as common in the C-suite of organisations I have spoken with as individual stories of workplace betrayal. I have come to realise that shutting down is proof that the theories explained in the book are correct.

However, I am optimistic because I have also engaged in conversation with people who can imagine better. I am regularly asked by friends, acquaintances, and colleagues what topics I research. When I mention moral injury, at first, I get a quizzical look, then a spark of interest, and then I'm often asked, 'tell me about that'.

After I explain the concepts, I hear, 'Oh yes, I could fill a book about *that* from my workplace…'. Then a light comes on as they recognise what was really going on for them at work. I hope you have such an 'aha moment' while reading, and I encourage you to fold a corner on the page when you do. The interactions I have with curious leaders, workers, and volunteers give me cautious optimism that things might change for the better as a new generation of leaders rises.

The goal of this book is to see 'better' managed virtue, understanding, staff well-being, and organisations. While 'better' may seem impossible, Snow, Smith, and Bazalgette provide us with a cautious optimism that change will occur with time. Let's get to the task of re-thinking, of learning a new language, and ultimately the work of removing the poison from our workplaces.

About the Author

Mark Layson (CESM MAIES) is Adjunct Research Fellow at Charles Sturt University, Australia, and the Director of the NSW/ACT Disaster Recovery Chaplaincy Network, placing pastoral carers in disaster zones. He serves as a volunteer Chaplain with the NSW Ambulance Aeromedical Unit, working alongside his therapy dog, Wallace. Mark's career spans multiple emergency services: 7 years as a NSW Police officer, professional firefighting, 18 years as an Anglican Pastor, and over a decade as an ambulance chaplain. His interdisciplinary PhD, funded by the NSW Centre for WHS, explored moral injury in Australian first responders, developing a holistic bio-psycho-social-spiritual framework for organisational culture change. A member of the Australasian Society for Traumatic Stress Studies and certified Emergency Service Manager, Mark's research focuses on wellbeing, psychosocial risk, and moral injury. He regularly presents at national and international conferences.

Part 1

Foundations of Workplace Betrayal

A true word.

> *There can be only one permanent revolution — a moral one; the regeneration of the inner man. How is this revolution to take place? Nobody knows how it will take place in humanity, but every man feels it clearly in himself. And yet in our world everybody thinks of changing humanity, and nobody thinks of changing himself.*
>
> Leo Tolstoy

As this book has evolved during research and writing, one significant problem vexed me. I've wondered if the issue of workplace betrayal and moral injury is simply a ruse. Is the concept of workplace betrayal merely a way to blame others for our own faults? Is calling out poor leadership and damaging cultures really a cover for one's inability to face personal demons? But in a casual conversation over coffee, I noticed a long-known reality had tacitly started to weave through the original transcripts as they took form.

While this book aims to bring about positive changes in organisational leadership, its more fundamental goal is to impact the heart and soul of each reader. Ultimately, our organisations and leaders are functions of each human working in them. The aim, therefore, is to initiate change within each reader, so that as they progress to leadership positions, those organisations will change. The route to changing oneself involves understanding both the systems in our world and our own selves.

The foundation of all knowledge is the ability to 'know thyself'. This reality has been evident since the ancient Greeks built the philosophical foundations of the

DOI: 10.1201/9781003499725-1

Western World, but as Tolstoy might agree, it is often neglected in the modern world. Therefore, our journey commences with self-knowledge. And with this knowledge, you can steadfastly refuse to lose yourself in the moral battles you will inevitably face in your workplace. By the end of Part 1, you may know yourself better, and only then can we see how the world of work might be revolutionised.

1 The Relationship between Identity, Meaning, and Work

A true word.

> *To love at all is to be vulnerable. Love anything and your heart will be wrung and possibly broken. If you want to make sure of keeping it intact you must give it to no one, not even an animal. Wrap it carefully round with hobbies and little luxuries; avoid all entanglements. Lock it up safe in the casket or coffin of your selfishness. But in that casket, safe, dark, motionless, airless, it will change. It will not be broken; it will become unbreakable, impenetrable, irredeemable. To love is to be vulnerable.*
>
> C.S. Lewis

INTRODUCTION: CHEESE PLATTERS AND POLITE CONVERSATIONS

People of my age throw middle-class, mid-life parties that have cheese platters, mid-volume 80's pop/rock, and awkward conversations with people with whom we had friendships many years ago. At one such party, I asked an old friend, 'John', an off-handed question, 'What are you up to at work?'. It is a question regularly asked at parties in the culture I inhabit. We had already shared pleasantries about the weather, sport, and the favourite topic in my hometown of Sydney – property prices. Running out of things to say, I casually asked this question. In retrospect, it was dangerous because, for too many people, answering it is difficult and even painful.

The question, 'what do you do for work?' may elicit delight. In the early days of one's working career, we can be doe-eyed and frankly chuffed not to be asked, 'what do you want to do when you grow up?' Admittedly, other young workers are restless, hoping for a better job or better pay. Those who love their work are frequently well paid or feel satisfied because their work makes a positive difference. Others get to work at something they are highly skilled at or enjoy doing. Finally, the most likely reason people enjoy work is that they have an esprit de corps with colleagues and leaders which produces a trusting workplace that is safe and friendly. However, for

DOI: 10.1201/9781003499725-2

many, talking about work can elicit a shame-tinged response stemming from a terrible job or boss, under-employment, or unemployment.

Most relevant to this book is the class of wounded conversation partner for whom work is now an ongoing drain on their sense of worth. Once an object of desire, even doe-eyed love, the job has now become a set of golden handcuffs or a form of slow torture. These workers can carry wounds that are not immediately obvious – wounds of the moral kind.

The ensuing pause from John was too long to be a thinking pause; it was soul-searching. It was accompanied by a long look into his half-empty beer, and one over each shoulder. After checking that the coast was clear, my old friend said in a hushed tone, '*Actually*… I'm looking for work'. This positively phrased statement, 'I'm looking…', was a euphemism to avoid saying the words associated with personal and societal failure – 'I'm unemployed'. Unemployment and underemployment have terrible consequences in our lives.

In 2013, a team of researchers in Melbourne, Australia investigated the impacts associated with a lack of meaningful work. Perhaps of little surprise, they found that the stigma associated with unemployment is associated with poor mental, social, and psychological health, including increased levels of suicidal behaviours. These negative impacts become more pronounced the longer one is unemployed (Milner et al., 2013). Loss of employment is truly high stakes.

Those who are fully employed are not entirely off the hook. Workplace stressors, including noxious workplace cultures, can lead to betrayal-type moral injuries. Unsurprisingly, these also have high correlations with poor mental health and increasing suicide rates (Fani et al., 2021). In summary, no work is inherently damaging to our sense of self; however, working in an environment where we are undermined and betrayed can be more destructive, as John's story would attest.

John told a zigzagging tale about the conditions at his last workplace. I listened and consoled John, in an attempt to both ease his pain and relieve my own guilt for asking a loaded question. As the conversation continued, details were revealed about how the last job had deteriorated over time, leading to an unhappy conclusion. He had been quickly paid out, while his concerns about the ethics of the business were left unattended. Indeed, much like Mike in the introduction, an unethical leader had been promoted while John was 'let go'. John's employment ended with a whimper marked by a brief and duplicitous companywide email noting his departure.

Our conversation also revealed a more profound shame than that arising from unemployment. Being betrayed by a company where he had done valuable work and which he had loved so much really hurt. What began as a much-loved job in a well-respected company became bad, even soul destroying. As C.S. Lewis noted, such is the ironic pain only the things we love can bring. Work is a double-edged sword in that we need it, and find value in it, but when we give ourselves to it wholeheartedly, we open the door to betrayal and moral injury.

I often have conversations about bad workplace experiences. In my doctoral research, 37% of people claimed to have been betrayed by managers, colleagues, or systems in their workplace more than 50 times during their career. I ask this question in seminar and conference presentations, where I have never received less than 25% of people claiming to have been betrayed more than 50 times during their careers.

I am not sure if hearing about the regularity of workplace betrayal is comforting because the pained are not alone, or alarming because it is so common. However, while these conversations are difficult, they are vitally important. At first glance, the confession 'I'm unemployed' is a hard social situation to avoid. Who wants to delve into that? I do, and it is my abiding conviction that only in the depths can we appreciate the heights.

I now research how to make sense of the stuff many people hope to leave unspoken. My doctoral research involved understanding and preventing betrayal in disaster and emergency workers by focusing on moral injury. Most noticeable in the literature was that much of the discussion on this topic related to extreme contexts. Until recently, moral injury and betrayal have only been researched in life-or-death occupations such as the military, policing, and medicine. In these contexts, the themes of betrayal and moral injury are sometimes conflated with traumatic events. The traumatic and extreme contexts in which moral injury is usually studied leaves most finding the topic scarcely relatable. While I also researched moral injury *in-extremis,*[1] my everyday conversations were with people who had everyday jobs, or no jobs – teachers, lawyers, accountants, social workers, pastors, scientists, vets, tradespeople, and others. Almost without exception, they understood when I explained my work and then wanted to tell me their stories of betrayal.

Difficult conversations about unemployment, betrayal, and trauma are vitally important human interactions. When we dare to step aside from self-interest and respond with empathy, the desire to listen, and some love for the other, it helps healing start. When we walk with wounded souls along their dark valley road, our presence and acceptance can move them towards healing. Being with wounded souls can also help the listener. Indeed, researching death, trauma, and betrayal has not led me to depression but an increased appreciation of life, a sense of peace, and increased faith.[2] You too may have an interest in difficult conversations about work. This chapter winds through the valley of betrayal and moral injury, but my fervent hope is that it will lead to positive outcomes for you, your loved ones, and the workplaces in which you labour.

This is deliberately a book about everyday work, rather than work with death or trauma. More specifically, it is about when the work we once loved turns nasty, as it too regularly does across the spectrum of careers. It will translate and expand the research on moral injury and betrayal from the language of academia and *in-extremis* to everyday language. The book unveils what lies underneath those rueful looks into our drinks and makes explicit a new vocabulary for what *actually* happened to you or any hapless party conversation partners.

Before we get to healing these moral wounds, we need to understand the reasons workplace betrayal is so impactful on human identity and meaning-making. Accordingly, the rest of this chapter is devoted to sketching the most basic outlines of the riddle that has long plagued humanity – questions of identity and meaning.

THE HUMAN SEARCH FOR IDENTITY AND MEANING

A Vital Survival Skill

Dolphins have innate cuteness and advanced faculties, including the ability to communicate via clicks and whistles. Dolphins can even collaborate in hunting for food. They are highly intelligent, and almost everyone adores them. Despite their sophistication, dolphins don't gather to discuss existential philosophy. Except for human beings, no other species pauses to consider ethics, epistemology, or metaphysics. Even among humans, only a small percentage pause to give explicit thought to their own existence and purpose. Many of us live an unexamined life, a kind of life Socrates reportedly said is not worth living – even with cheese platters and '80s pop/rock!

Like an iceberg's tip, the tiny percentage of a whole, many consider our existence and the possibility of transcendence only fleetingly. In fair weather, this approach to meaning and identity mostly works as we sail through the calm waters of life. However, when seas turn rough through the winds of betrayal and moral injury, we can become shipwrecked by the enormous mass of our unexamined philosophical and moral icebergs. It is in hardship that we can find ourselves without the tools to understand, let alone repair, the depth of damage betrayal has had on our sense of self. My doctoral supervisor reckoned many of us fail to exercise our moral imagination to understand our inner convictions (see Chapter 9) before hardships come. If we haven't understood ourselves and our moral world, we may be left scrambling to solve the damage to ethical structures we don't even know the depth of (Cameron, 2011; Johnson, 1993).

A complete examination of the meaning of life is not possible or feasible in this book; however, a brief outline of what these questions look like is vital. Australian military ethicist Deane-Peter Baker has said that humans are better able to cope with morally challenging environments when we understand who we are, by considering questions such as 'What do I really believe? Who am I? Where do I belong? What is my purpose?' (Cox, 2018).

The task of exploring our identity and meaning, therefore, is not a practice of navel gazing, but a necessary workplace survival skill. Research regularly reports that betrayal-type moral injuries significantly impact mental, physical, social, and spiritual health. Moral injury is associated with increased suicidality (Fani et al., 2021; Jamieson et al., 2023), loss of moral virtue (Atuel et al., 2021; Shay, 1994), and decreased workplace performance (Layson & de Almeida Neto, 2023; Mehnert, 2022). However, those who have taken more than a cursory examination of their moral code and meaning in life appear to have an extra layer of protection from the morally injurious impacts of betrayal (Blumberg et al., 2022; Kelley et al., 2021).

We will undertake the vital task of outlining human identity and human meaning with a slightly artificial split between identity and meaning-making. Both concepts are like two strands of DNA with interlocking personal units of meaning and identity. To begin, let's make a simple observation that identity is about *who and what* we are, while meaning is about *why* we are.

Identity: Who Are We?

Humans are complex beings who have attributes unmatched by any other creature in our world – these are not just *quantitatively* unparalleled, whereby humans have a higher level of achievement in the same qualities that other creatures possess. Humans are *qualitatively* unique, and this makes us uniquely vulnerable to betrayal and moral injury. Like a well-cut diamond, humans are multi-faceted creatures that have various elements that make us what and who we are. We can appreciate that each of these facets is true without denying that the others are simultaneously true. Indeed, we lessen the glory of the human diamond when we deny any of them.

Humans Are Physical Beings

As you sit reading this book in a comfy lounge, lying in bed, or perhaps listening to it while you commute, the most obvious answer to who or what you are is that you are a physical being. You can easily note that you have mass; after all, the sofa or bed dips with your weight and other commuters must make space for you. You have hands to hold your book or tablet, and you have fingers to turn pages. When you arrive at your destination, you will require a physical desk or a station to work at, or some other physical accommodation that makes allowance for your existence. Humans, like dolphins and other animals, are predominantly a mix of oxygen, carbon, hydrogen, and nitrogen. Solely accepting the physical facet of human identity is what Harold Noonan, in a brief survey of thinking about human identity, calls the 'bodily criterion'.

Noonan states, 'to this view, personal identity is essentially no different from the identity of material objects… like a ship, or a living thing, like an oak tree or a horse' (Noonan, 2019, p. 2) – I assume cute dolphins too. The bodily criterion can seem like an attractive position to hold, as it means humans need not obey moral rules we don't like and which we feel impinge on our individual freedom. After all, ships and trees don't have moral laws to follow.

With a bodily criterion of humanity, we will be as free as a swimming dolphin. However, a purely physical view of humanity, freed from moral constraints, has consequences – we become part of the food chain. We are forced to submit to those who physically dominate and mistreat others. Under this criterion, your feelings of workplace injustice and betrayal are figments of your imagination, not real injustices, and you are not morally injured – just weak!

The German philosopher Friedrich Nietzsche led the way with this mechanistic view of a liberated humanity. Nietzsche argued that morality was invented by weaklings who lacked alternative means to control strong oppressors, so they created moral rules and punishments for those who breached them. The theory goes that rules about fairness, equity, and the kind treatment of the weak should all be regarded as fiction created by those with a slave-oriented morality. An unadorned bodily criterion leaves no hope for justice, which may be too high a price to pay for liberty from morals. We may start to think that the bodily criterion of identity pulls up short as an explanation of who or what humans are at their core.

Humans Are Thinking Beings

You would not be alone in feeling a little unsatisfied with the bodily criterion of identity. As obvious as this appears, other problems arise in physicality being the terminal point of our identity. The 17th-century philosopher and mathematician René Descartes introduced a difficult question to one-eyed supporters: what if we only *appear* to physically exist? This may seem an odd question to ask, but if you have seen the movie *The Matrix* you may get a sense of how one could be sceptical. In *The Matrix,* humans are kept unconscious in liquid-filled caskets so that they can provide energy to their super-intelligent computer lord. The computer projects a fake physicality into the minds of all humans, making physical reality a mere dream.

Well before the dystopic scenes in *The Matrix*, Descartes entertained radical doubt about everything, even physical existence. He leveraged doubt about everything to dig deep and find something he could not doubt, to have a point of certainty from which to build human identity and meaning. He noted he was thinking about questions of identity and meaning at the most basic level. If he was thinking about it, then he must at least exist in a cognitive sense. Hence, 'I think, therefore I am'. Humans, therefore, are more, or maybe less, than carbon-based creatures. To Descartes, humans are thinking beings that happen to have the secondary experience of physical existence.

Another problem addressed by a cognitive understanding of human identity is the distinctiveness of higher-order reasoning. Ships, oak trees, horses, and dolphins are unlike you in that they will not buy this book. Other objects in our world don't buy or read books. Human cognitive abilities are not merely *quantitatively* greater than other carbon-based objects; our mental abilities are *qualitatively* different – they are unique.

Humans uniquely can ponder, wonder, philosophise, psychologise, and even theologise. That is, humans can intimately and robustly consider the immanent conditions of life as well as the transcendent realities of eternal space and time. Therefore, our identity is more than a mere physicality; it requires us to think of ourselves as thinking beings. We all do that to some extent. You think about work, you think about betrayal, you think about wounding and healing.

Behind all this thinking is not just a carbon-based bodily organ that we call the brain. Necessary for this function is what philosophy and psychology both call the human mind. The abstract reasoning of humans serves as a foundational assumption in both the art of philosophy and the science of psychology. If we agree that humans think in their mind (not just their brain), we will accept a two-faceted view of human identity that could be called a bio-psychological view of human identity. At this dualistic moment, we face a point of decision in our attempt to answer our questions about human identity.

Firstly, we can hold to the bodily criterion and reject the notion of humans as unique thinking beings. However, I would find it somewhat ironic to hold bodily criterion and buy this, or any, book. Secondly, we could accept that Descartes' thinking creature hypothesis sufficiently explains who we are and reject the bodily criterion to embrace *The Matrix*. Thirdly, we could adopt a dualistic view, believing that both physical and cognitive realities explain human identity, and then proceed to the next chapter. Finally, as the protagonist in *The Matrix* does, we can 'take the red pill' and explore other possible dimensions of human identity.

Humans Are Collaborative Working Beings

If you haven't skipped to the next chapter already, I assume you took the red pill and are interested in exploring even more depth to human identity. A good place to start is the level most relevant to our current interest, the identity of humans as workers. Humans are not merely embodied minds that contemplate existence. More importantly, humans don't just *do* work; we are workers who utilise physical and cognitive abilities to work at a fundamentally existential level. When we recognise our identity as workers, we can appreciate with new clarity the depth of impact from workplace betrayal. Despite possible objections that work is merely a consequence of physical existence and cognitive ability, it is essential to resist a trend to separate what we do from who we are.

We live in an age in which people increasingly seek, often with good intent, to separate who we are from what we do. Many years ago, I was rightly chided for describing someone by reference to their medical diagnosis. To describe someone as an 'epileptic' is to pigeonhole them with a label and strip away essential elements of their identity. Instead, the more accepted description for such a person is 'someone who suffers from epilepsy'. Such sensitivity is important; however, separating ontology (our being) from praxis (what we do) can be taken too far. A person may extend this logic to say, 'I am not a thief; I am a person who suffers (and often profits) from stealing things!'

We are right to question attempts to completely separate the inseparable parts of what we are and what we do. For many generations, the truth of being what we do was so obvious that we named families after their chief occupation, for example, Baker, Smith, Cooper, Priest, or the straight-shooting Fletchers. At a time when towns in England were geographically isolated but growing, occupational specialisation became more commonplace. It is worth noting that in history, an obvious answer to someone's identity was their work. The importance of work to community cohesion and an individual's identity in that community is so central to our understanding of workplace betrayal that it bears repeating – as humans, we do not merely work; we *are* workers.

The issue of humans *being* workers comes home with some gravity for those who are 'in between jobs', 'looking for work', or more bluntly 'unemployed'. Given our regular complaints about the drudgery and stress of work, surely a break from work would have at least a tinge of pleasure. However, the chance to spend more time with family, sleep in, and explore hobbies (as well as attend to undone chores) rarely seems to compensate for not having communal and/or paid work to do.

On a personal note, the essence of being a worker came home to me like never before at the conclusion of my PhD study. Throughout my adult life, I'd have had a long series of continual work and study since leaving school. There had never been a time when I was uncertain of the work I did or who I worked for. But then, when I was most qualified, I found myself unemployed for several months. It was soul-crushing and made me consider again why work is so central to who we are.

Allow me to advance two parallel theories about work and meaning. Firstly, in evolutionary terms, work has been central to our existence. In tribal life, if a person refused to work, it endangered the existence of the entire group. Those who refused to work lost the ability to belong in the group and as a result to procreate. In this way, working became intimately connected to physical and social existence in our own generation and any succeeding ones.

Secondly, in the various creation narratives throughout history, such as the one found in the biblical book of Genesis, work is not just a coincidence. Like the God who is reported to have undertaken the work of creation, work is who we are in our being. The Genesis account says humans were created not for leisure, or for pleasure, but to work. Wherever one lies on the spectrum between evolutionists and creationists, work is central to who we are. And yet now that work is too often broken, it is painful and burdensome despite being vital to our sense of worth. Our relationship with work is somewhat like Gollum's relationship with the ring in Tolkien's *The Lord of the Rings*. Gollum both loves and hates his precious! We, too, may hate our work only at a level exceeded by our need for meaningful and constructive work.

Humans are also a special kind of physical, thinking, working being. We are increasingly inventive and necessarily collaborative. Humans are not just workers; our work creates social networks, and the benefits of our cooperation have a recursive relationship with communal societies. Humans both specialise and cooperate in social groups for mutual benefit. For example, a town full of bakers might fill our stomachs but leave our horses without shoes. It is at the point at which we recognise humans as workers that we start to move beyond individualistic categories of human identity.

Humans are social beings. With limited exceptions, human beings gather in interdependent communities as a basic expression of who we are. Humans don't just have individual physical, cognitive, or work-related existence; we exist in an interconnected web of ever-increasingly complex societies. These societies start with families but then expand to tribes, villages, cities, nations, and finally to all humanity. Humans do not merely think by themselves – they create schools, universities, and communities of academics. Similarly, humans work and develop tools within organisations, companies, and governments that benefit us and the societies we live in.

What we have described thus far can be summarised as biopsychosocial identity, made popular in the 1970s by psychiatrist George Engel (Engel, 1977). Whilst Engel's focus was to understand the cause and treatment of disease better, his work was an important contemporary summary of the nature of humanity. Dissatisfied with the dominant biomedical model reducing humanity to physical/biological processes, Engel convincingly demanded we take a broader view of the human condition.

By the start of the 21st century, the biopsychosocial framework was the dominant framework through which humans were studied. As important as this was, it stopped short of acknowledging a final element that broadens our existence beyond the humdrum of everyday life. This final element of human identity, that we are spiritual beings, becomes the bridge to understanding the meaning in humanity – the 'why we are' questions.

Humans Are Spiritual Beings

Perhaps it was the cultural milieu of late modernism or a caution about rushing into territory where angels fear to tread, but Engel and many of his contemporaries largely ignored the spiritual element of human identity. For as long as humans have considered their physical, cognitive, and social states (also known as *immanence),* they have both suspected and experienced something that transcends that reality. That *transcendence* relates to the spiritual element of our identity. Like most things of ancient provenance, the spiritual component of human identity had been tied to religious expressions of transcendence. However, as the 21st century progresses, there is an increasing recognition that spirituality is a related but separate element of the human state that now even allows people to call themselves 'spiritual but not religious'.

Spirituality is not strictly linked to transcendence, at least not etymologically. In Jewish and Christian texts where spirituality is critical terminology in the original languages, life itself is in view when spirit is mentioned. The word spirit in Hebrew (*ruach*) and Greek (*pneuma*) refers to the simplicity of breath/wind/air. To be inspired is to be breathed into. The Genesis account details how God breathed His spirit/breath into the first human so that they may come to life. The author of Genesis never imagined this breath was purely physical. It was a holy breath – a Holy Spirit that animated human existence. Following this divine inspiration into humanity was a sense that a mark of 'the other' perpetually existed in humans. As spiritual beings, we developed beliefs about the afterlife and the significance of the physical tasks we undertake in this life, including our vocation.

An example of inspirited work may include the motivation to work in roles that feed the hungry, assist the poor, and commune with the lonely. Nietzsche's embrace of the bodily criterion abhors such a slave morality approach to life and work in which the weak are assisted. Kind and altruistic acts are not simply tasks one may, or may not, choose to do as an amoral act. Charitable work has been infused with moral gravitas because it is *the right thing* to do.

Even beyond altruistic work, non-charitable works are often seen as acts of spiritual significance. Great composers such as Bach and Handel inscribed their musical scores with the letters SDG – *Soli Deo Gloria* – To God alone be the glory! They saw their earthly works filled with significance when dedicated to transcendent realities. More mundanely, the 16th-century reformer, Martin Luther, is often reported to have held that the cobbler partook in something of transcendent significance because, when done well, shoemaking was a chance for the cobbler to love his fellow human.

Finally, of immense importance to both spirituality and moral injury, you might note that makers of shoes and doers of charity have two ways of doing those tasks. Any profession has a technically *right* way to make things, such as shoes, and secondly, a morally *right* way to make shoes, show kindness, or do any productive work. For example, shoe-making is a specific skill; however, the morally right way should not be ignored. Is it the right thing for a cobbler to make sturdy shoes for Nazis? Is it right for shoemakers to make their shoes last only a week beyond their warranty? Humans produce guidelines about the technically right way to do things, but they also create laws, ethics, and morals about the morally right way.

Humans are natural rule makers. As sociologist Christian Smith writes, humans are moral believing animals that always create and inhabit moral systems (Smith, 2003). Unlike all other beings, humans praise and chide actions as either morally good or bad rather than on the simple criterion of utility. Indeed, Nietzsche noted we're not morally vexed by eagles preying on lambs – it is what eagles do. Nietzsche, therefore, objected to any moral positioning about humans preying on other humans – it is what strong humans should do! However, with hindsight, we have reasonable cause to object to the callousness of Nietzsche's solution as much as any 'nazified' interpretation of his work that led our world into the Second World War (Whyte, 2008). Without arguing the point further, it is proposed that humans have a transcendent spiritual and moral facet. This spirituality may or may not involve belief in God or religious adherence. Regardless, our spiritual element profoundly impacts our meaning-making and moral sensibilities.

When the spiritual element is included alongside other parts of human identity, the resultant – and now holistic – picture is that humans maintain a biopsychosocial-spiritual (BPSS) identity. When we detail the impacts of moral injury shortly, the relevance of this well-rounded framework will become more evident. But it is worth noting that these elements interact with each other in a dynamic way to create the human soul.

The word soul (*psyche*) has long associations with the spiritual facet of human identity and is the etymological root of psychology. However, it is better deployed to describe the overlapping elements of humans. As researcher and friend Kate Bradford writes, we do not so much have a soul, but we are each a soul (Bradford, 2023). When the captain of the Titanic gave the SOS (Save our Souls) he wasn't asking for spiritual salvation, but salvation from physical death.

At the end of examining human identity, you are entitled to ask, 'So what if we're multifaceted souls; does anything we do *mean* anything in the big scheme of things?' In response, we must immerse ourselves in the idea of meaning-making.

Meaning: Why Are We?

What, if anything, survives after we die? Does anything we do truly count? Will there be more than a headstone with our name on it – perhaps a hospital wing or university? In life, is there any connection between events and people other than mere physical causation, such as genetic connections within a family? Or, when applied to work, is there anything about our work, or the way we undertake it, that extends beyond rearranging atoms before we die?

These are all hard questions, but to channel Descartes somewhat, we know that most humans are asking questions about meaning. Meaning can be summed up as the human drive to find connection and significance in events and the possible relationships between them. Social psychologist Roy Baumeister writes 'meaning connects things' (Baumeister & Landau, 2018). These connections and significance are the base of the iceberg that many humans neglect to examine until times get tough.

My work has brought me near many who are suffering as a result of betrayal, illness, and death. Many have not considered the meaning of their life until difficult times force them to consider it. Usually, those who have considered meaning-making less develop more tortured answers under the duress of suffering, and those who have examined the iceberg of suffering beforehand often do much better.

Neurologist and psychiatrist Viktor Frankl was probably the 20th century's most important writer on meaning. Frankl, an Austrian of Jewish heritage, survived while his first wife, parents, and brother were killed in Auschwitz and other concentration camps. His experiences of suffering have made him the perfect scholar on meaning. In two groundbreaking books, *Man's Search for Meaning*, and *Man's Search for Ultimate Meaning*, he alerts us that searching for and finding meaning in the worst of places is not only possible, but vital for physical, psychological, and spiritual survival.

Many who hold to a bodily criterion of identity do so in an attempt to be free from the binds of morality; however, this kind of freedom places them in an ocean void of meaning. Nietzsche and others may obtain some sense of freedom but ultimately find themselves carried by the tides and currents of moral fashion or the moment into an ocean they are unable to navigate. Thus, nihilists and reductionists find themselves more lost than they are free. Frankl was certain it was the spiritual component giving meaning and purpose that transcends the physical.

Inner freedom, even in-spired freedom, provides something that can never be taken away, whether in a concentration camp or a university campus. When a holistic BPSS framework is conceded, it allows bodily, psychological and spiritual freedom in a world full of meaning. We need to navigate the complexity of the holistic freedom we seek, and this book will help us understand and navigate some of that complexity. I have gone to great lengths to reach holistic identity and meaning, because righting ourselves after being capsized by workplace betrayal and moral injury is impossible with simplistic or monochromatic views of human identity.

Human meaning-making processes involve all elements of the BPSS identity, and the biological and physical aspects of work are essential to consider. For example, if we are overworked to the point of burnout or are relegated to menial tasks, it can be hard to derive meaning. Therefore, it is vitally important that our workplaces enable and empower us to help create physically and psychologically safe workplaces. Furthermore, our cognitive functioning as it relates to self-efficacy and accurate self-appraisal will deeply impact our ability to make meaning. Workplace components such as leadership, morale, and social cohesion are elements that can make or break our ability to make meaning. Finally, our spiritual drive to create moral systems and connect our work to transcendent themes plays a central role in our soul search for meaning.

With the language of moral injury and workplace betrayal, you can develop tactics and a solid moral framework before adverse events happen and avoid being lost in a meaningless world and difficult party conversations. Understanding our connection between identity and work, and nurturing meaning-making structures, is central to unlocking positive responses to betrayal and moral injury in the modern workplace.

EPILOGUE: THE MODERN WORKPLACE

The modern workplace is not what it used to be. Work has evolved from tribal hunting and gathering to small villages and then into large cities. In tribes, working was a matter of survival. In villages, humans settled down to create safe places to live, specialising in communal work, while in the modern city, we are increasingly alienated from our local communities and reliant on work to provide meaning and identity.

As religiosity declined across the Western World over the 20th century, mechanisms to make meaning shifted from church and community to the workplace. People sought work less for survival and more as a source of meaning. Despite the massive expectation for workplaces to create meaning and identity, the laborious nature of work and fractured relationships mean workplaces are unable to fulfil the enormous burden to provide existential meaning to its workers.

Urbanised workers are now divided, both loving and hating work and seeking meaningful self-actualisation from something we detest. We increasingly invest more of our moral expectations into work and expect more than pay for our work – we demand meaning from it. Our desire to obtain meaning and identity from a job is unrealistic; instead, work can be a place in which we express our meaning and identity. This distinction requires us to seek other avenues of meaning and identity so that even the worst of workplaces cannot destroy us. Nonetheless, occupational dissonance is sapping, and in many ways, this dissonance lies at the heart of moral injury and betrayal in the workplace.

ITEMS FOR CONSIDERATION

- What does it mean to have overlapping biological, psychological, social, and spiritual elements to my identity? How might I nurture each of these to prevent distress?
- Am I putting too much of my search for self-worth onto the workplace?
- How might I connect my search for identity and meaning to holistically engage with work and community life?
- Am I at risk of moral injury by having greater moral expectations of work and its leaders than they can possibly deliver?

NOTES

1 *In-extremis* is a Latin term used to describe the most difficulty and deadly situations humans face.

2 These elements of growth are part of what is known as Posttraumatic Growth, and which we will discuss in Chapter 9.

2 Dissonance
What Else Were You Expecting?

A true word.

> *... moral injury involves an act of transgression that creates dissonance and conflict because it violates assumptions and beliefs about right and wrong and personal goodness. How this dissonance or conflict is reconciled is one of the key determinants of injury.*
>
> (Litz et al., 2009, p.698)

INTRODUCTION: WHAT ARE YOU EXPECTING?

Several years ago, I became socially and occupationally well acquainted with a man of prestige and significant influence in my sphere of workplace operations. Not only did Tom have a PhD, it came from what is widely regarded as the best university in the world! With a sharp mind and tremendous work ethic, Tom had risen near the top of his field. During that time, I attended a conference with a close friend, which, by coincidence, Tom was also attending. I was excited to introduce my friend to Tom, hoping it would benefit my friend's career and inspire us both. My wife and I invited Tom, my friend, and their wives to share a meal. They say it's a dangerous thing to meet your heroes, but as I was to discover, just as dangerous to introduce them to your friends.

Tom did nothing but complain about almost every aspect of the conference in the most demeaning of ways, and at the end of the meal, I was somewhat in shock. After Tom left, I felt compelled to apologise to my friend. I knew Tom had a reputation for bold criticism, but this was unexpected, next-level poison. I had hoped, given our relationship, his intelligence, and position of influence, that he would be a sociable meal companion. I never shared my expectations for that meal, neither before nor since that day. Instead, these expectations were my little secret. I left that meal with the conviction that, as I progressed in my life and career, I wanted to be nothing like Tom, and I vowed never to introduce him to my friends.

Tom did not break any rules or laws, but he did break my expectations of polite, friendly, and intelligent conversation, which I had thought were modest and reasonable expectations. A law does not have to be broken for moral injury to occur; in fact,

DOI: 10.1201/9781003499725-3

when rules are broken, there is the possibility of an independent arbiter deciding appropriate recourse. But when expectations are broken, we can experience a profound dissonance, with no avenue to right the wrong. While my dinner fiasco with Tom did not cause a moral injury, it contradicted my expectations and caused a dent in my worldview.

Although this was somewhat of a trivial event, it helps illustrate the power of broken expectations. When unmet expectations involve matters of substance such as the security (and even sanctity) of our career, personal and professional reputation, deepest beliefs, and our meaning-making senses, then moral injury can occur. Workplace moral injuries arise from substantial dissonance between our deeply held moral expectations about fair workplace behaviour, regulations, and relationships. When co-workers and leaders transgress our expectations and then threaten our livelihoods to cover it over, few go unscathed.

This chapter details the relationship between three kinds of dissonance behind moral injury. The first is dissonance at the relational level when we have poorly aligned expectations of the depth of relationships with our organisation or its key leaders. Secondly, and related, is dissonance between the explicitly stated ethical commitments of an organisation and our deeply held moral convictions. The third is the difference between contractual obligation and a covenantal faithfulness that our moral expectations desire. The idea of covenantal faithfulness is then compared to psychological contracts about our work, which, when dishonoured, may lead to moral injury. Having laid an expansive theoretical framework throughout Chapters 1 and 2, we will be in a good place to conclude with a full explanation and definition of moral injury.

DOMAINS OF DISSONANCE

Thick and Thin Relationships

There you are, near the front of the line, after waiting for over 30 minutes. Suddenly, a stranger asks if he can cut in ahead of you. He tells you his wife is heavily pregnant, and that the 30 minutes would mean an uncomfortable wait. With some reluctance and begrudging agreement from your fellow queue sitters, you let them in. But about an hour later, you see the queue-cutter walking hand in hand with an athletic-looking woman who is definitely not pregnant!

Let's change that story slightly – instead of a stranger cutting in, your former college roommate shows up telling you her husband is terribly ill with gastroenteritis, and it would be challenging for her to wait in line while he sits nearby trying not to vomit. You happily let your friend in and even have a quick catch-up. However, about an hour later, you see her and her husband downing a milkshake, burger, and fries at the local burger joint.

What emotions and thoughts are stirred within you? Both the stranger and your friend have deceived you equally, but perhaps one action angers you more than the other. More than likely, you would consider having a trusted friend lie to your face as the greater insult. What makes a difference is the depth of the relationship you hold with them. You had different levels of relationship with the fraudsters, which in turn impacts your reaction to the insult.

Philosopher Avishai Margalit, in his book *On Betrayal* (Margalit, 2017), describes this as the difference between *thick* and *thin* relationships. Thick relationships are those with whom we maintain the strongest bonds of mutual service, friendship, or love. On the other hand, thin relationships are the kind we have with casual acquaintances, or with strangers we pass on the street each day.

These are not categorical labels – instead, each relationship exists on a spectrum from thin to thick. It is highly personal where the lines might be drawn between soul mate, trusted ally, reliable team member, colleague, suspicious acquaintance, or devil incarnate. And the truth is, we have work colleagues and bosses who fit all of these.

It is safe to say the relationship with the stranger in our scenario was thin. Because of this, the pain of deceit is real but only penetrates the outer layers of our expectations and worldview. A thick relationship would have developed over the years with our college roommate, so the insult challenges our beliefs about friendship, trust, and perhaps even the reliability of moral order. This depth of impact, indeed betrayal itself, can only be caused by those with whom we have thick relationships.

Margalit notes the irony of thick friendships by quoting the adage 'where there is no love, there is no betrayal'. Betrayal-type moral injury in the workplace arises from unexpected dissonant interactions with those at the thick end of the relational spectrum – a trusted colleague, a seemingly noble leader, or when trust in our own judgement and behaviour is violated. Moral injury was first recognised in military contexts because the thickest of relationships are regularly forged in the crucible of trauma and suffering – the bond a soldier forms with comrades is ironclad. For others, battling together against a bad boss or difficult client builds thick relationships. When close workplace relationships are combined with the desire for meaningful work, the risk of moral injury increases.

As we recognise the dangers associated with high expectations of colleagues and workplaces, the language of morality becomes appropriate. Margalit further explains something obvious about the varying thickness of relationships: humans have different moral frameworks, which vary according to the thickness of our relationships. He states that these different codes are the difference between morals and ethics.

Thick and Thin Rules

Segregating *morals* and *ethics* is a fraught activity, but worthwhile. Since historically they have been used interchangeably, one danger of exploiting this nuance is that we could read meaning into the words that their original authors never intended. Any distinction between morals and ethics should be applied cautiously; however, the division does have some utility. In leveraging the semantic nuances of morals and ethics, there is an opportunity to understand better the discordant personal and organisational codes that create the dissonance inherent in moral injury.

Margalit helpfully makes this distinction, asserting that ethics applies to the domain of thick relationships, while morals guide our thin relationships. There is great merit in Margalit's proposal; however, I believe the labels need to be turned 180 degrees. I suggest that morals regulate thick relationships, while ethics are the stuff of thin relationships, and I make this alteration for two reasons.

The first reason is due to the nuances of the current usage of the terms. If I were to tell you that a nefarious act you committed in the workplace is 'unethical', you may be taken aback. Fortunately, the well-meaning folks in HR have created a corporate ethics statement poster that can help adjudicate my allegation. If I let the poster tell us what is ethical or not, some form of external arbitration on the claim of unethical behaviour is possible. In our dispute, I might specify that your act did not 'empower' me in any way; therefore, what you did was unethical.

On the other hand, if I said, 'what you did was immoral', there would be no poster to help adjudicate my claim. Instead, we would each hold very different views. What my claim of immorality lacks in objectivity is made up for by the depth of emotion arising from the breach of those deeply held moral expectations. Morals cut to the core of our meaning-making structures, whilst modern-day ethics are often presented in memorable acronyms, and are the stuff of sanitised and performative corporate expectations.

The second reason is that moral injury results from the breaching of our deeply held moral expectations, which are more subjective than the ethical statements that are often presented in a glossy manner. Australian military ethicist Deane-Peter Baker has written that moral injury in military personnel is primarily caused by the 'critical gap between the moral identity of the individual who dons the uniform in service of the state [...] and the ethics [...] of the state' (Baker, 2020). The space between a worker's moral expectations and an organisation's *thin liberal ethics* is a petri dish of dissonance and confusion where moral injury thrives.

Destructive dissonance occurs because humans are not detached, acronym-loving ethical spectators; we have moral flesh in the game. Sociologist Christian Smith writes that humans are moral, believing animals whose deeply held beliefs inform our morals. Moral psychologist Jonathan Haidt goes some way to explain how our moral sense forms through his Moral Foundations Theory. Haidt writes that humans don't form their moral frameworks through rational debate or workplace committees, but rather through the interplay of family life, social interactions, cognitions, lived experiences, and ideological frameworks. Our moral expectations often remain tacit because they are usually a 'vibe' that many people fail to examine thoroughly.

Furthermore, Haidt claims that humans are more than moral; we are moralistic, critical, and judgemental (Haidt, 2013). That is, we form harsh judgements about the motives and character of those with whom we disagree. Judging from the heated and increasingly polarised political debates of our day, it is hard to argue with Haidt's assessment of moralism and judgementalism. When someone who we thought shared our moral expectations acts against those convictions, it can be perceived as betrayal, leading to a sense of moral outrage. This level of outrage cannot be tamed by organisations promoting sanitised and democratised ethics statements. Corporate statements are insufficient arbiters of moral truth.

Morals

Having made such large claims about morals, we must define what is meant by them. Most research publications on moral injury, with the notable exception of Litz et al. (2009), define moral injury but rarely *moral*. Building on that of Litz et al. and influenced by Jonathan Haidt's Moral Foundations Theory as expounded in *The Righteous Mind*, I define morals as:

> The personally held fundamental assumptions about how things should work and how one should behave in the world. Morals are formed intuitively and recursively in the context of shared familial, cultural, societal, religious, and legal rules for social behaviour and are often held tacitly. Their content transcends explicit or institutional codes of practice, and while often overlapping, can be dissonant from explicitly stated organisational ethical frameworks.

Our moral framework is a core part of each person, and like our fingerprints, it is unique to each of us. So, it is nearly impossible to find a perfect moral match between the closest of friends, let alone with the ethical standpoints of organisations. As a result, some moral and ethical dissonance is to be expected in every morally capable person – to live is to be morally affronted. Being challenged in our moral frameworks is healthy for the humble, so the aim of this book is not to prevent moral dissonance but to stop it from damaging our sense of self to a level that degrades our moral virtue.

Ethics

As social and moral creatures, humans benefit from explicit guidelines on how to treat each other fairly and equitably, and so we have written down those expectations in various formats. The resultant ethical and legal statements are necessary but insufficient for healthy moral interaction. For large communities, such as countries, humans make ethical statements through legislation, while in smaller contexts, specific legal contracts often include a code of ethics. Such agreements are documented and may include things such as which side of the road to drive on, outlining financial remuneration for work or services, employment contracts, and even regulating personal relationships through marriage certification.

These binding agreements are explicitly stated and made known so they can be applied in most scenarios and with the thinnest of relationships. Because of their general application, our laws cannot completely deal with every circumstance, nor can they deal with our individual moral fingerprints. An example of explicit ethical statements not entirely aligning with everyone's moral code can be found in marriage agreements. When two individuals marry, they make legal promises that include that they are neither currently married nor closely related to each other.

The happy couple also verbally promise that they will not commence an intimate relationship with another person until death parts them. Despite this, there is considerable variance in each partner's moral expectations, bringing expectations from their family of origin and personal preferences. For example, one loves to snuggle while the other is physically distant; or one has an obvious delineation for what constitutes an intimate relationship with a third party, while the other may have a flexible definition of extra-marital sexual relations, as Chapter 6 will explore.

It is the passage of time that thickens this relationship and fully reveals the moral expectations each party has brought to this arrangement. It is vital to have legally binding arrangements as both a limit on poor behaviour and a protection for vulnerable parties. While signing legal marriage certifications is important, legalistic adherence to them does not guarantee an emotionally, physically, mentally, or spiritually satisfying relationship. Satisfaction in these areas is built on deeper moral virtues that include grace, love, truth, faithfulness, forbearance, and forgiveness. While these virtues make a terrible ethical acronym (GLTFFF), they make for enriching personal and workplace relationships.

A workplace example of the insufficiency of written codes comes from the world of medicine. The Four Pillars of Bioethics contain guiding principles on which modern Western medicine is based. As robust as these principles appear, they can be impotent in times of personal tragedy. Intensive care doctor, Ken Hillman, writes,

> I reflected on the irrelevance of the four principles of beneficence, nonmaleficence, autonomy and distributive justice. They are of little help when considering withdrawing and withholding treatment, futility, dying, heroic measures, explanations, transparency, honesty and compassion.
>
> *Hillman, 2010, p. 168*

Whilst neither Hillman nor I are advocating for the abandonment of explicit ethical statements, I am asserting in the strongest terms that corporations have become what George Eliot called 'men of maxims', distilling life's mysteries into general rules. In the 21st century, organisational ethics are often facile maxims put up on a poster, not the stuff of personal conviction and belief. Ethics statements are primarily decided by committees, approved by legal, and packaged by media departments into pithy maxims that manage thin relationships but cannot change behaviour nor speak to our souls.

It is better by far to take a holistic BPSS framework (see Chapter 1) as the foundation of our need to find identity and meaning in work, because the legalism of exterior workplace codes of ethics is insufficient to maintain human meaning. While workplace ethics have utility, our moral expectations must be explored, for the mysteries of human relationships profoundly shape them. When we synthesise thick/thin relationships with morals/ethics, we will start to create a language to explain how we might better relate in life, love, and workplace settings.

Smooth or Textured Relationships

Contracts

Your written employment contract outlines the organisation's expectations, which likely include being productive, honest, and reliable. In return, your employer will pay you, provide a workspace, and offer some level of protection from harm. Once you and the senior leader sign the contract and you've done a 1-hour online module on the all-important code of ethics, you're ready to go.

Explicit contracts are like the cold, hard surface of an industrial kitchen. Such hard contractual benchtops are great for getting the job done, but they're terrible places to eat. Contracts are necessary, but are cold, hard insurance policies that apply when trust is low among strangers or enemies. Despite their necessity, such bare minimum standards are relatively impotent when more personal and implicit expectations are brought to the table. For example, I made no formal contract with Tom about his behaviour before we shared a meal, but I maintained a deeply embedded expectation of how it *could* and *should* have happened. I was not relieved that Tom hadn't broken a written or verbal contract, but was disappointed to have my own expectations dashed.

Contractual relationships sit at the thin end of the relational thickness spectrum, with ethics, laws, and codes of practice. However, workplaces have moral, meaning-making humans in them who need something more than cold contracts to feel safe, productive, and happy. Therefore, we must think of better ways of guiding workplace relationships that transcend contractual obligations. Another, more textured way of describing relationships is that of a covenant.

Covenants

Covenants are textured, nuanced, even bespoke manifestations of the deep human need for relationship, which contracts fail to satisfy. Covenants are more like a well-set table at which we can enjoy a shared meal that was cooked on that cold, hard steel bench. Covenants have functionally unnecessary elements, much like a dining table has candles, flowers, and tablecloths. Such a setting is where you want to have interconnected workplace relationships. There is still a need for the workbench (contract), but the profound need for the textured dining table of human fellowship is better expressed by covenantal thinking. Workplaces that lack covenantal behaviours are like restaurants that produce excellent meals, but are hard to enjoy because of poor service and ambience. Contracts and covenants are both required in productive and healthy workplaces.

The kinds of covenants I speak of do not negate contracts, nor oppose them; instead, covenants transcend contracts. In legal circles, the word covenant is regularly used to define a particular kind of contract, especially in property law. However, this current legal usage has obscured traditional covenantal thinking, which is steeped in ancient promises and practices that guided community, belief structures, and interpersonal relationships. Such older understandings of covenants have several helpful aspects worth applying in the workplace.

Simply defined, a covenant is an 'agreement between two parties binding them mutually to undertakings on each other's behalf' (Archer, 1999). At first, this may not seem very different from contracts, but several components of traditionally conceived covenants push the limits of Archer's definition. Firstly, contracts and covenants have different 'spatial' qualities. Contracts create a lower limit on behaviour; they prevent the worst-case scenario. One is not allowed to go below these standards, such as pay and workplace conditions, without fearing legal recourse. Alternatively, covenants are an open sky that allow for aspirational thinking about how workplace relationships and behaviour can be mutually beneficial in the space above contractual compliance.

An example of legal lower limits versus covenantal upper possibilities is seen in marriage. The signatures declaring one is not currently married nor related to your spouse-to-be form a contract that sets no specific aspirations. However, the promise given to love one another 'in sickness and in health' is aspirational covenant thinking. It looks positively towards what could be possible if unconditional mutual love is expressed and experienced.

An example of covenant thinking in a workplace is found in military and policing organisations. Soldiers and police officers all sign legally binding employment contracts; however, a regularly implicit, and often explicit, promise is made that the recruit has not just started a job but joined a family. It is what some in policing circles call a 'blue family', or in the military, a 'band of brothers'. This talk of familial care and commitment is reminiscent of a covenant mindset. Familial type expectations, even if unspoken, influence a recruit's perception of what is possible in this new family. Away from the extremes of policing and the military, many who work in philanthropic organisations or ones with altruistic and caring intent imagine they will work in a way that a contract does not capture.

A second difference between covenants and contracts stems from the commitment to a set of values that transcend the relationship. Contracts are largely transactional, and their ethical components apply to a specific context, such as the workplace. However, covenants are inherently relational and aspire to more than minimum standards. In a covenant, the parties commit to each other and to ideals that transcend the relationship. Contracts centre on tasks and obligations, while covenants speak outside of an agreement to a *way of being*. In many cultures throughout history, the proper moral *way of being* was through faithfulness to covenantal ideals over simple obedience to the letter of the law.

Covenantal faithfulness is particularly relevant to building better workplaces that mitigate the risk of betrayal. Covenants contain the aspirations that organisations and their leaders project, such as familial care or social justice, so they do more than fulfil a contract. Covenantal faithfulness is a two-way street with increased moral expectations being placed on workers and their leaders. Such faithfulness is vital for workers when they perceive they have been betrayed and are tempted to resort to illegal or destructive acts of retribution (Chapter 9). A covenant asks a betrayed person, who may be morally injured, to continue their commitment to the aspirations in the covenant and thus maintain personal integrity.

As an important aside, the notion of covenantal faithfulness is a check and balance on the dark side of loyalty. Before researching betrayal, I assumed loyalty was the cure to betrayal. While I still value loyalty, I now do so with caveats. I came to understand that even a cursory glance at loyalty reveals its use by mafia syndicates or corrupt public officials to manipulate people into complicity with their immoral conduct. Loyalty demands preferencing relationship over moral principles. Many a dark deed has been committed in the name of loyalty.

One final difference between covenants and contracts is that covenants may be entered into unilaterally and tacitly. A contract always makes public, enforceable claims on both parties. A positive example of a unilateral covenant is the love of a parent for their child. In the birthing suite, a parent often pledges themselves to the child's wellbeing, regardless of their understanding of the child's potential characteristics. No initial demands are made – it is a 'no matter what' covenant. A parent working on the basis of covenantal faithfulness, even when disciplining their child, does so in accord with a promise for the child's good.

A difficulty with unilateral covenants in the workplace arises when relational expectations of a worker do not align with realistic workplace relationships. When an employee creates an unrealistic covenant about what they are owed, they are likely to face significant disappointment. Many of us enter the workplace with an internal moral covenant about the workplace. Another way to understand this is through the lens of psychological contract theory.

Psychological Contracts: Better Still – Moral Covenants

Cindy had a successful business career, but something just hadn't clicked in the roles she held. Her job in marketing and media with a large multinational was fine, but it didn't feel like it truly contributed to the greater good. Cindy had a strong urge to 'do something worthwhile', so with a little trepidation and a downward shift in income, Cindy resigned so she could take up a role with a non-government organisation (NGO) she admired. Cindy's desire to serve, combined with the NGO's amazing projects for disadvantaged people, made it a perfect match. Cindy figured that with morally good work, a morally upright organisation, and similarly minded colleagues, it was going to be like a family.

After making some great connections with colleagues over the first year, she settled into a regular pattern at work. But something slowly started to feel wrong about the more experienced marketing manager in another region. She had met Roger while doing a field visit, and like everyone else, she was very impressed and perhaps a little star-struck. But there was a high turnover of staff in Roger's region. His team were producing great work, but they were also churning through staff at a rapid rate. Then, one of Cindy's close friends in Roger's office suddenly resigned. When Cindy contacted her to check all was okay, her friend reluctantly told a story of intimidation and verbal and emotional abuse by Roger towards clients.

While Cindy was alarmed at these revelations, she believed her organisation's lofty code of ethics would help address the issue, so at the next opportunity she approached her local manager. The next day, Cindy's director emailed her with a calendar invite. At the meeting, Cindy thought she would be able to raise her concerns and start a plan to have them addressed. After telling what she now knew to be a clear pattern of poor managerial behaviour, her director paused and said, 'Yes, I do know about this, but they really kick goals across the organisation. We can achieve more good with Roger than without him. Surely you wouldn't want to put our work back, would you?'

Cindy's manager promised to have a quick word with Roger, but suggested that she concentrate on her region and keep this confidential. 'After all, you can imagine the damage this would do to our work and our brand if it got out'. It was never the same again. The moral dissonance was too large. Cindy lasted another five months and left with a broken heart.

Mike's story at the start was unitary, while Cindy's is an amalgam of different stories people have told me over coffee or a beer. Cindy's story makes my interaction with Tom seem feeble. However, what Cindy and I shared was an unmet expectation. Like Cindy, we can all be dazzled by the good reputation of a philanthropic organisation, believing our experience would be positive and morally coherent with its goals and our expectations.

When a person commences a new job, they create what psychologist Denise Rousseau labelled a Psychological Contract. When we sign a written workplace contract, the terms are transactional, objective, and enforceable, while the moral freight and tacit obligations of psychological contracts are highly subjective. Rousseau notes that what a written contract means 'is a subjective experience for each participant, affected by personal values and upbringing, the relationship's history, and the broader society. All of these factors merge to create a central feature of employment: the psychological contract' (Rousseau & Schalk, 2000).

Rousseau described two kinds of contracts: psychological contracts and implied contracts. While both have relevance to this book, we will be concentrating on psychological contracts in this discussion. The reason for focusing on psychological contracts is that her description of what fuels a psychological contract mirrors elements of our moral framework. Individuals' personal values and deeply held moral expectations are both highly subjective, morally laden, and unilaterally declared (Rousseau, 1989). Importantly, if our unwritten expectations are not met, they constitute a breach of our psychological contract, and more acutely, our deeply held moral expectations.

While Rousseau's work is profound and accurate to the experience of workers across all industries and cultures, I would argue that an adapted title of *moral covenants* may better describe the content of these psychological contracts.

Firstly, moral covenants better describe the arrangements we unilaterally make about our workplace expectations. The identity of humans drives our expectations as moral, meaning-makers. The drivers of these covenants are our deeply held moral expectations and not *psychological* ones. Rousseau herself wrote that the psychological contracts are 'affected by personal values and upbringing' but 'affected' is far too weak, I fear.

I contend that psychological contracts are fundamental expressions of our deep meaning-making desires and human moral frameworks. These moral frameworks, as Haidt rightly asserts, are not strictly rational or psychological; they are intuitive. Alternatively, as we will read in Chapter 9, moral reasoning requires a level of *imagination* to understand and describe our framework. To label intuitive and/or imaginative moral expectations as *psychological* leaves open the possibility that our moral expectations are rationally formed, which they are not.

Secondly, considering the previous discussion on covenants, it would be a stretch to describe one's inward moral expectations about work as a transactional contract. Contracts are ground-level insurance policies that are explicit and enforceable. In contrast, our moral covenants are aspirational and contain relational expectations that transcend the transactional arrangements of a contract.

This relabelling of psychological contracts as moral covenants is more than simple pedantry; it is vital for the intimate relationship that psychological contract theory has with betrayal and moral injury. I had not considered this until I heard about the groundbreaking work of Jennifer O'Neill-George, who is researching this connection in the Australian Defence Force (ADF) (O'neill-George, 2022). The connection between our moral expectations and betrayal is more than just philosophical – it lends real power to understanding the depth of hurt when our moral covenants are leveraged at recruitment and weaponised during our careers. It turns out many organisations have little will or ability to fulfil these empty covenantal promises.

Weaponising Moral Covenants

Rousseau argues that psychological contracts or *moral covenants*, as we have labelled them, are strongest when an organisation makes a promise of good treatment to an individual worker. Therefore, at recruitment and during initial training, any promises of familial care or the use of noble means to achieve noble ends will set up strong and hopeful moral expectations in the individual's mind, heart, and soul. Manifestations of these expectations include the hope of promotion or increased remuneration in exchange for hard work (an implied contract), but more so the expectation of organisational adherence with moral goods, including justice and truthfulness.

In 2008, the satirical political comedy *The Hollowmen* dramatised the faceless bureaucrats in the Australian Federal Government. In an episode called *Military Matters*, the plot surrounded the need to recruit more members for the ADF. In addition to focus groups, the episode featured consultations with key stakeholders. In one consultation, the faceless bureaucrats asked senior military figures what images should be shown in the recruitment campaign. To meet recruiting targets, the ranking military officer asked that no dangerous or life-threatening situations be depicted; 'It's a real negative'. The response surprised the bureaucrats and made a viewer shake their head in disbelief that such folly might occur even in fictional settings.

But within the decade, life was imitating art as the ADF ran a recruitment campaign weaponising moral covenants. The campaign was entitled 'Do What You Love'. If you love travel, being with friends and keeping fit, the ADF is for you. The promise of a lifestyle career was made to prospective workers, but it came at the expense of being truthful about the need for sacrifice and courage in the face of danger. Not surprisingly, this campaign did not solve recruitment shortfalls, and on reflection, it was noted that the campaign sent 'a lifestyle message to potential recruits rather than targeting the specific capabilities that the ADF needs' (Australian Army Research Centre, 2022). Furthermore, the campaign failed to reflect the reality of life in a hierarchical fighting force, where one frequently does not do what they love but must do as they are commanded! How can one's moral covenant not be in danger when aspirations are weaponised in this manner?

At the time of writing, the ADF recruitment campaign had changed its recruitment tag line to 'Make your impact' in an attempt to leverage people's moral expectations of doing good. Much like the lure Cindy was hooked by at the start of this chapter, the tag line aims to leverage our meaning-making desire to 'do something worthwhile' in a job that is 'like no other'. In the post-pandemic labour shortage, similar campaigns urge people to do something worthwhile across many careers.

People are regularly encouraged to join emergency services, the medical profession, and even publicly listed corporations that highlight their environmental, social, and governance (ESG) credentials. They all aim to tap into recruits' moral expectations. Leveraging the desire to do something worthwhile taps into the urge we discussed in Chapter 1 about our need to find meaning in work. Organisations often make duplicitous claims, rarely discussing contractual obligations or the enforcement of employment contracts in the event of an error or illness.

The duplicitous weaponising of moral covenants during recruitment helps forge unrealistic moral expectations that may have profound long-term implications. In my 25 years of experience in and around high-stress professions, I have seen that breaches of moral covenants relating to reward for effort, provision of care, and organisational justice cause far more distress than less frequent breaches of written contracts. Research is also increasingly showing that these moral stressors are more significant in the formation of burnout and mental health problems than exposure to death and trauma.

Described above is a precondition of the experience of betrayal in the workplace. The betrayal occurs when an employee is given an unrealistic picture of a role, leading to a potent and weaponised moral covenant. If the reader thinks the above description of moral covenants is accurate, it is worth readers considering the difference between their own moral covenant and their signed employment contract. Furthermore, if you perceive a strong dissonance, be cautious about exacting acts of vengeance because of this betrayal.

You may have developed a belief that the organisation is making a covenant that honours your deepest moral expectations; however, when the rubber hits the road, the organisation is likely to drop to the level of the contractual compliance floor. When you are compelled to do monotonous paperwork, in remote locations away from family at the command of a mildly sociopathic leader, you may say, 'I didn't sign up for this!'. In your moral covenant, you signed up for travel, fitness, friends, and the chance to do something worthwhile! However, the response often comes, 'yes, you did, here is your signature'. And in this honest or imagined debate about what one signed up for, we find the dissonance that leads to moral injury.

MORAL INJURY

As commander of an elite military unit, Andrew was revered by his soldiers, feared by enemies, and relied upon by his country. Despite his prodigious ability to make war, Andrew regularly showed restraint against enemies and received many military honours from his epic victories. But after the most significant battle victory of his career, and in a possible case of stolen valour, a rival commander falsely took the commendation Andrew had earned. The rival commander not only usurped him but also did so in front of Andrew's subordinates and with the knowledge of his superiors. As a result, Andrew was both outraged and shamed.

After this incident, Andrew withdrew from frontline duties and did not see combat for many years. The overwhelming sense of betrayal Andrew felt started to skew his view of himself and the service he had rendered. But whilst on this self-imposed hiatus from battle, his closest friend was killed in action. In a combination of grief and outrage, Andrew reengaged in combat duties. But this time, he was different. Gone was his former restraint; Andrew now killed indiscriminately, and, lowering his moral code further, he even desecrated the bodies of slain foes. The starting point of Andrew's demise was betrayal, its trigger was grief, and its end was a complete loss of identity and moral ability to act as an upright warrior.

Andrew's story is actually an old one. Andrew does not exist, and the account above is a brief retelling of the mythical life of Achilles, who was depicted in Homer's epic, *The Iliad*. The account of Achilles' betrayal and downfall was central in the thinking of Jonathan Shay, the founding father of current conceptions of moral injury. Shay worked for many years with Vietnam Veterans and detected strong resonances between their experiences and the emotions and destructive rage of Achilles.

Not only did Shay appreciate the Greek Classics, but he was also a classic rethinker. In the decade and a half after PTSD became an official diagnosis in the Diagnostic and Statistical Manual (DSM), Shay was rethinking the miasmic trauma-focused consensus about the causes of distress in war veterans. Much like the cholera outbreaks in 19th-century London were blamed on an evil miasma, the poor mental health of veterans was almost exclusively blamed on exposure to traumatic events. However, after years of treating Vietnam war veterans, Shay wrote in *Achilles in Vietnam*, 'I do not believe the official PTSD criteria capture the devastation of mental life after severe combat trauma' (Shay, 1994, p. 169). While the PTSD diagnosis took the traumatic *content* of wartime experiences seriously, it largely ignored the moral and organisational *context* in which those traumatic events occurred.

Shay's realisation about the importance of context in the formation of distress arose in parallel with nascent models of workplace stress mitigation. Not long after Shay wrote Achilles in Vietnam, a team of Work Health and Safety researchers, led by Tom Cox, independently made the distinction between the *content* and *context* of workplace stress formation. Among other things, Cox and his team highlighted the damage caused by contextual stressors such as 'poor relationships with superiors, interpersonal conflict, and lack of social support' (Cox et al., 2000, p. 4) – faint, less *extremis,* echoes of the leadership betrayal Shay presented.

Much like Snow's theories about the waterborne nature of cholera transmission, the importance of workplace context in the formation of distress was largely unheeded for a quarter of a century. Contextual mental stressors are only now being addressed through workplace safety standards known as psychosocial hazards, which we will return to in Chapter 10. Still, it is worth noting that the same tension between content and context runs through workplace stress mitigation and the various ways of defining moral injury.

Defining Moral Injury: Context

When I teach, I am adamant that my students define their terms carefully because, as Led Zeppelin said in *Stairway to Heaven*, 'you know sometimes words have two meanings'. Much against these convictions, I have so far failed to define either of the key terms in the title of this book, but that is about to change. As we start our definitional section with moral injury, another element of Zeppelin's oracle lives. Moral injury has at least two meanings, derived from two foundational definitions. In the introduction, we briefly noted the first definition in Jonathan Shay's oft-quoted summary of moral injury as,

a. the betrayal of what is right,
b. by someone in legitimate authority,
c. in a high-stakes situation.

Shay puts so much emphasis on context that the above tripartite definition is barely a definition in any traditional or clinical sense. It mentions nothing about symptoms or impacts, as one might find in the DSM when it defines and describes disorders. Shay did speak of symptoms and effects of moral injury at other points, including rage and degradation of character. Still, his emphasis, at what many consider the definitional level, is purely contextual.

Shay emphasised the organisational context, particularly the kind of leadership betrayal Achilles experienced. More mildly, Cox labelled such contextual stressors as poor relationships with superiors and interpersonal conflict. The modern aphorism 'people don't leave a bad job; they leave a bad boss' speaks to this reality. In Shay's view of moral injury, the traumatic content of combat is not irrelevant, but it's not the main game. For combat veterans or the lawyer/teacher/academic who has uttered in frustration that they have left a bad boss, contextual stressors are key.

Saying Shay's definition is not really a definition does not lessen its importance in my mind; on the contrary, it heightens the importance of his stance. What Shay highlights is the missing key to rethink the failure of wellbeing programmes. Shay and Cox's insights offer an opportunity to reduce workplace distress at a population level. While many workplace wellbeing strategies reactively help individuals, Shay opens the door to stop the incidents that cause that stress and distress in the first place.

Shay, Cox, and others enable primary prevention strategies because they allow us to prevent the preventable. For a combat veteran or first responder, the traumatic content of work is not a modifiable variable. Bad leadership and poor organisational support are preventable in any workplace if leaders of our organisations have the will to do so. Shay's non-definition highlights the need to target workplace strategies that prevent moral injury related to leadership factors.

Defining Moral Injury: Content

Over the years, research has progressed our understanding of moral injury and highlighted its different facets. In 2009, psychiatrist Brett Litz and his team of fellow researchers developed a preeminent definition of moral injury, stating it to be 'the lasting psychological, biological, spiritual, behavioural, and social impact of perpetrating, failing to prevent, or bearing witness to acts that transgress deeply held moral beliefs and expectations' (Litz et al., 2009, p. 697). Litz and his team did three important things.

Firstly, they brought the *content* of morally injurious events back into play. This definition brings us back to a concept similar to PTSD, that is, the morally contentious content of a person's work is the causal factor of distress. Much like trauma exposure, morally contentious work is the stock and trade of the soldier or the first responder. Morally contentious decisions include whether to kill an underage armed combatant, whether to return a child on a court order to a drug addict step-parent, or whether to inflict a painful medical treatment on a patient for whom it is unlikely to improve their viability. Morally contentious work is also imaginable in civilian work undertaken by the vet, the teacher, the lawyer, or almost any profession. If one takes a minimalistic view of Litz's definition, then moral injury is just a slightly altered version of PTSD that is, by and large, beyond the scope of the organisation to prevent.

Secondly, their conception of moral injury brings the individual worker as the primary agent to blame in stress formation. In Litz et al.'s definition, the individual themselves may have perpetrated or failed to prevent a potentially morally injurious event. That is, people who do the wrong thing or fail to do the right thing, and those who are witnesses, are front and centre. These two types of individually focused moral injuries have been further divided by some into self-directed moral injuries (when we do the morally wrong thing ourselves) and other-directed moral injuries (when others commit a morally questionable act that we fail to prevent) (Currier et al., 2018). With both Shay's and Litz's definitions in play, three flavours of moral injury are notable: 1) betrayal, 2) self-focused, and 3) other-focused moral injury.

Now ask yourself a question: if you were an HR leader or a lawyer who defends organisations from claims of harm, would you celebrate Shay's version of moral injury that implicates leadership betrayal and organisational culture as the cause of moral injury? Alternatively, would you lean into Litz et al.'s individual-focused perpetration model? With Shay, senior and middle management would be required to take responsibility for morally dubious actions they fail to prevent and morally unsafe cultures they breed. If a simplistic application of Litz et al. wins the day, blame for distress can be placed back on those suffering distress. On the dissonance one may experience about these questions, Australian military chaplains and wellbeing researchers Tim Hodgson and Lindsay Carey make a strong point in this extended quote.

> Researchers in developing their own definition of moral injury have tended to placate leadership malpractice and/or the role of the chain of command in causing moral injury to veterans. These researchers have commandeered Shay's term "moral injury" and refocused its scope away from those of "legitimate authority" to focusing specifically upon the behaviour of the individual—who may or may not be impacted by leadership or organizational decisions. Whether deliberate or unintentional, diminishing the responsibility of organizational leadership, can subsequently minimalize and undervalue the sense of betrayal perceived by current and former serving personnel and effectively exonerated corporate culpability and the liability of organizations with respect to moral injury.
>
> *Hodgson & Carey, 2017*

The answer to resolving this dissonance in Shay's *context* thesis and Litz's *content* antithesis is to synthesise these approaches. To its credit, the ADF has adopted just such a synthesised definition of moral injury, which includes the physical, psychological, social, and spiritual impact of grievous moral transgressions, by self, others, or betrayal by those in authority (ADF, 2021).

It is this holistic understanding of moral injury that will be used as a referent point throughout this work; however, the focus here is on betrayal-type moral injury, for good reason. As moral injury research increases, it is often reported that distress related to betrayal-type moral injury has higher associations with suicidal behaviours and poor mental health outcomes than other types of moral injury (Greenberg, 2022; Jamieson et al., 2023). Secondly, I am entirely convinced that, if correctly identified, acts perceived as betrayal by workers can be significantly curtailed if organisations have the will to remove them. That is, betrayal is both the most dangerous form of moral injury and the most preventable for people like Mike, Cindy, and you in your workplace.

The third significant change Litz and his fellow researchers introduced was the holistic nature of harm inflicted by moral injuries. It is striking that these researchers addressed the entire range of human identity discussed in Chapter 1. Moral injury is not strictly a psychological phenomenon, a religious model, or a social malady. Moral injuries are socially, behaviourally, biologically, and spiritually devastating phenomena. The reason we spent so long unpacking the bio-psycho-social-spiritual nature of human identity is that moral injury impacts them all. Unless one is willing to understand the existence and importance of all these domains, helping those who carry moral injuries will be a puzzle that is missing pieces.

I hope this book inspires interdisciplinary teams to set aside any disciplinary rivalries and differences, listen to each other, and collaborate to put all the pieces together. Collaboration is central to preventing the moral injuries that are impacting our workplaces. Indeed, so far, we have referred to the research of anthropologists, chaplains, demographers, philosophers, psychologists, psychiatrists, social workers, theologians, and work health and safety professionals. I point this out in the hope that all these professions and more will engage collaboratively to help fix broken workplaces. Without this team, we cannot solve the curse of moral injury that is caused by betrayal.

ITEMS FOR CONSIDERATION

- What are the enticements your organisation is currently using to recruit staff? Are they realistic, and do they form any part of the written contract with recruits?
- What do you currently consider to be the most stressful parts of your current job? Are these contextual or content factors within your role?

3 Betrayal
The Personal Side of Dissonance

A true word.

> ***A riddle:***
> *Q. What is a method of attack an enemy can never inflict upon you?*
> *A. Betrayal! 'Where there is no love, there is no betrayal.'*

INTRODUCTION: BROKEN PROMISES AND A BROKEN HEART

Alice and Stephen met at high school, fell in love, and married young. By the time I knew them, they had been married for over 20 years and were raising two teenage children. Life was busy. Like many their age, life involved two demanding jobs, community volunteering, and driving kids around the city for ever-increasing commitments. Like most marriages, theirs had slipped into a certain predictable routine. The early passion of their relationship had evolved into a consistent partnership that gave certainty and regularity in the busyness of life. Stephen travelled regularly for work, while Alice juggled her job and day-to-day life at home. There were no significant difficulties in their marriage: no violence, illness, financial problems, and no warning of what was ahead.

One day, the certainty and regularity of their partnership were shattered when Stephen announced, out of the blue, that the marriage was over. Stephen had been having a series of adulterous relationships that now held more excitement than the marriage. Alice played back all the times she believed Stephen was working hard on business trips but was really philandering. She felt foolish to have been deceived so badly and was experiencing total heartbreak that the sweetheart of her youth had been so two-faced. When Alice told me, she said, 'I just feel totally betrayed'.

Those we trust the most have a unique ability to harm us in ways our enemies never can. Betrayal comes from close friends, allies, lovers, and those who are close enough to inflict a deep wound. With greater levels of trust comes a thickness of relationship that brings closer physical and psychological proximity. It is this proximity

DOI: 10.1201/9781003499725-4

that allows the knife of betrayal to be plunged deep. Betrayal in the workplace comes from those with whom we build the strongest relationships of trust and camaraderie, or from the leaders we rightfully expect honesty and integrity. But what are the other ingredients of betrayal?

BETRAYAL

Linguistically, the essence of betrayal centres on the idea of movement from a place of safety to a place of uncertainty. Betrayal resembles being thrown into the ocean by a force that is so large and proximal that you cannot resist. In ancient Hebrew, the root word for betrayal originates from the word for throwing something. So too, in ancient Greek, the word for betrayal, *paradidōmi,* is used for the mundane process of handing over an object or knowledge from one person to another. The English word 'betray' finds its etymological roots in the Latin word for handing over. At its most benign, the word involves passing something to another. At its most malignant, it is used to describe the transition from trusting relationships to relational unfaithfulness and broken social trust.

More technically, betrayal has been defined as 'a voluntary violation of mutually known pivotal expectations of the trustor by the trusted party (trustee), which has the potential to threaten the well-being of the trustor' (Elangovan & Shapiro, 1998, p. 548). Key to this definition of betrayal are five key elements:

1. a voluntary act that involves,
2. a person's pivotal expectations,
3. that are mutually known,
4. then violated,
5. with the potential to cause harm.

This technical definition lacks the nuance that has come with newer research on moral injury. With the research on psychological contracts (moral covenants) and moral injury, the expectations (Point 2) that are betrayed do not have to be mutually known to cause harm. Moral covenant breach is perhaps the most crucial kind, as it is a betrayal of our deepest-held beliefs about work.

The above definition also lacks a deeper understanding of the harm caused by betrayal. At an experiential level, betrayal vigorously throws an individual into a dark and turbulent ocean of self-doubt and broken social trust. Betrayal has a significant impact on the identity and meaning-making structures we discussed in Chapter 1. When betrayal occurs, it attacks a central part of our meaning-making structures, making it hard to know what, if anything, is certain anymore.

Betrayal in our personal and professional lives is not merely a corporate contract violation that, while harmful, is actionable through legal means. Betrayal is a breach of our deeply held moral expectations and resultant moral covenants and brings a real and existential threat to our sense of identity and meaning. Given the depth of damage betrayal causes, it is no surprise that the act has been seen as worthy of the highest condemnation throughout human history.

Betrayal in History: The Very Pit of Hell

In Chapter 2, we heard the Ancient Greek epic of Achilles' betrayal by another commander – the seminal act that provoked Achilles to lose all sense of his own virtue and restraint. While Achilles' fall is a central theme of *The Iliad*, and his betrayer plays a minor role, other historical accounts give greater emphasis to the actions and fate of those who betray others.

Can you imagine what it would be like to travel to the very pit of hell? What would you see? Which villains and crimes would be on display? How would you grade the severity of punishment for one offence over another? Where would betrayers be positioned? The 14th-century Italian poet and moral philosopher, *Dante Alighieri*, described such a magical and disturbing journey.

Dante gave a detailed account of being led on a journey to the heart of evil in the famous work, *Divine Comedy*. While Dante would eventually tour heaven, he first travelled ever downwards through the nine 'circles' of hell, the solemn journey commencing with a warning to 'Abandon all hope, you who enter here'. His trip to hell and back is both disturbing and informative.

Dante wrote of a journey down through ever increasingly grave evils in each circle of hell. In the shallowest portions of hell were those guilty of lower-grade wrongdoing, such as lust, greed, anger, and even violence. When Dante finally enters the 9th, and final, circle of hell, we are shown the gravest sin of all. Dante pictures the centre of hell as surprisingly frozen and still, rather than the chaotic fires many might imagine, and it is reserved for the worst offenders – betrayers!

In the centre of hell were those who betrayed family and guests in their homes, and those who were politically duplicitous. However, right at the centre of hell, Satan himself was consuming the most heinous betrayers in history – Brutus, Cassius, and Judas – infamous characters who betrayed their masters, Julius Caesar and Jesus Christ, respectively.

In Dante's worldview, betraying those of lower or equal social stature was heinous, but betraying one's own masters was definitively evil. It is this observation that betrayal also travels up the hierarchical ladder, a factor regularly overlooked in moral injury, must delay us for a short time. Betrayal is not only committed by leaders against their subordinates; it works up and down social and hierarchical scales. It is worth considering what actions each of us may have played in undermining a leader through gossip, disrespect, and even lies. Many leaders I speak to feel a keen sense of betrayal by those they lead.

Returning to Dante's vision, there in icy punishment at the centre of hell, resided all those who breached familial, social, national, religious, and political allegiances. Notably, at each level, the betrayal was intimately personal. As he was being stabbed in the back by his close counsel and friend, Caesar cried out to his trusted confidante, '*E tu, Brute?*' – '*You too,* Brutus?' Caesar may have expected death at the hands of others, but death via Brutus' final dagger thrust was an unfathomable betrayal.

Judas, the other character in Dante's 9th circle, has become shorthand for betrayal – to call someone a Judas is the ultimate insult. When folk music legend Bob Dylan first plugged in an electric guitar in public instead of his acoustic, one of his biggest fans yelled 'Judas!'. Did Dylan really betray that crowd and his adoring fans? Probably not, but that loud detractor certainly felt betrayed, and perception is often king. Chapter 9 describes strategies to determine if something is truly betrayal or a poorly considered perception. The nature of Judas' betrayal was even more intimate than Brutus' because it was achieved with a kiss. Judas' signal to the lynch mob was ironically to give a kiss on the cheek to his master. Betrayal, it seems, comes in varied guises that include kisses, daggers, and broken expectations.

Philosopher Avishai Margalit outlines three traditional categories of betrayal that roughly align with those found in Dante's inner circle of hell. Margalit proposes that three categories of betrayal have been widely recognised through history: adultery, apostasy, and treason (Margalit, 2017). Respectively, these are betrayals against family, against community (particularly faith-based), and political betrayal. But it is worth asking if these three still resonate with 21st-century sensibilities. The first two categories have become somewhat mundane in Western society; however, people like Alice would tell us, they still sting at a personal level. Western cultures are vacillating on the third category, treason.

Scholars debate whether Brutus was a traitor or a patriot, much like today with Wikileaks founder Julian Assange, seen as a treasonous betrayer or a whistleblower who exposed immorality. Individual perspectives agree that betrayal is profound but argue about what kinds of acts manifest it. While betrayal's exact boundaries are hard to define, we can be sure it will harm our sense of being. The harm caused by betrayal will be examined shortly; however, we must draw in more tightly to explore the contours of how betrayal is experienced.

Operationalising and Quantifying Betrayal in the Workplace

> I shall not today attempt further to define the kinds of material I understand to be embraced within that shorthand description… But I know it when I see it.

So ruled Justice Potter Stewart in the 1964 case of *Jacobellis v. Ohio*. Stewart was not trying to describe betrayal but pornography, believing he intuitively knew what it was, should he stumble across it. Similarly, we might feel betrayal is hard to define but easy to recognise. Nonetheless, later courts tightened the outlines of how to recognise pornography so that objective judgements could be made. Similarly, recent research, including my own, is working through ways to *operationalise* and *quantify* the existence and experience of workplace betrayal.

In my doctoral research, I navigated this double-barrelled process through several steps. My immediate context was emergency workers, including police, fire, paramedics, and emergency medical clinicians. I collected 21 biographies of these workers, read them, and re-read them, making notes along the way about the stressors these workers faced. Then 16 of these emergency workers agreed to be interviewed so that I could dig deeper into what their experience of workplace betrayal might look like. These interactions allowed me to tighten the focus on the kinds of events and workplace interactions that may cause moral pain. So, what types of events did they perceive as potentially morally injurious?

Firstly, I identified two *content-type* moral stressors. These stressors were inherent in the nature of emergency work. They included death, rather than trauma, and particular kinds of deaths – the death of 'innocents' such as young children and women at the hands of violent forces. For example, when a morbidly obese 60-year-old man dies of a heart attack, it is sad, but not unexpected for emergency workers. However, should a child die as a result of a similar condition, it would be deeply impactful because it would challenge the worldview about innocence and suffering.

Related to the death of innocents, the other thing that morally pained these workers was the existence of injustice. They wondered how bad people could literally get away with murder, while innocent children could be ordered to return to drug-dealing guardians. These stressors are real and serious, but mainly out of the control of their employers. The genuine interest I had was to discover the *contextual* stressors that could be managed and mitigated. That is, I sought to identify what elements of the workplace cause a sense of moral pain from broken expectations.

From the large body of data I had, four main themes were perceived as betrayal and abandonment of workers by their organisation. The four themes are:

1. Organisational injustice.
2. Poor leadership support.
3. Excessive workload and under-resourcing.
4. Poorly applied reward and recognition.

Organisational Injustice

As we mentioned in Chapter 2, many workplaces today have published policies and codes of ethics that speak in generic terms about fairness. They promote concepts like transparency, accountability, and inclusion, but when it really counts, these same workplaces operate very differently. The stories of Mike from the introduction and Cindy from Chapter 2 make this dissonance plain. For many workers I spoke with, especially in first responder organisations, fairness felt like a performance, in which virtue was signalled rather than lived.

Some described the lack of transparency in transfer or promotion decisions, made without their input and presented as 'for their own good'. One person was informed that they were being relocated under a staff rotation policy for their wellbeing – yet no one consulted them, assessed them, or provided care. Ultimately, they were transferred quietly during the Christmas shutdown period. The so-called system of empowerment and consultation didn't feel at all like care – it was more like command-and-control leadership poorly applied.

Others spoke about how their mental health struggles were treated with suspicion or even used against them. One of my research participants described being encouraged to go outside and exercise to support their recovery – only to find that photos of them doing that were used to undermine their insurance claim. Instead of support, they were met with disbelief and what felt like punishment for being honest about their vulnerabilities.

These incidents of injustice are not isolated to emergency organisations. They point to a deeper cultural issue: a way of managing people that outwardly adheres to ethical standards and employment contracts but violates weaponised moral covenants. Some researchers call this 'administrative violence' (Wadham & Connor, 2022). Such violence occurs when policies and procedures, while technically legitimate, are used in ways that shame, belittle, or punish employees. These forms of injustice maintain a form of plausible deniability about the harm being done, but their impact causes injuries of the moral kind.

When systems become punitive rather than restorative, emphasise blame over learning, and where staff fear being open about their struggles, it is often perceived as a betrayal where moral injury becomes more likely. Restoring a sense of justice means more than tweaking policies or ethics statements; it requires the kind of significant cultural change discussed in Part 3 – a culture that recognises the human cost of unfair processes and seeks to rebuild trust and proper accountability through compassion, restoration, and continual improvement.

Poor Leadership Support

The protracted battles of the Afghanistan military campaign left a new generation of soldiers with physical and psychological wounds. It was assumed that the brutality of the battles had left the scars. However, British researchers reported that mental health was impacted more by leadership, unit cohesion, and overall morale than exposure to combat (Jones et al., 2012), suggesting that leadership is one of the most potent forces acting on worker distress in any workplace. Leadership shapes the tone, sets the standard, and can either buffer staff from harm or contribute directly to it. In the stories I gathered, leadership was often the turning point – either towards healing or hurt.

When leadership went wrong, damage was significant. Some reported senior staff who were physically or emotionally intimidating, or who pressured them to act unethically and immorally. Others described leaders who ignored them completely, dismissing their input because of lower rank or lack of status. This authoritarian leadership can leave employees feeling powerless, voiceless, and betrayed.

Here is an essential factor we must consider about leaders who behave poorly – they may be acting out of their own moral injuries. In his second book, Shay describes two styles of leadership that Achilles or Odysseus drifted towards after their woes. Their actions of attack or deception serve as a signpost about poor leadership by the betrayed,

> *When social trust is destroyed, it is replaced by the settled expectancy of harm, exploitation, and humiliation from others. With this expectancy, there are few options: strike first; withdraw and isolate oneself from others (e.g., Achilles); or create deceptions, distractions, false identities, and narratives to spoil the aim of what is expected (e.g., Odysseus).*
>
> *Shay, 2002*

Underlying the attack or deceptive leadership styles lies a profound sense of broken trust. Poor leaders are often broken and in need of support and restoration rather than punishment.

While unsupportive leaders both suffer and inflict betrayal, there were bright spots in what participants said of them. Many participants spoke warmly about their immediate supervisors – leaders who listened, showed up physically and emotionally, and treated their team members as people first and task-doers second. These leaders often worked on a first-name basis, took time to check in, and led by example. Their presence made a huge difference. Supportive leadership, participants said, is not about being perfect but about being real. Leadership that does not morally wound involves listening, being present, and having the courage to show vulnerability. One of the most helpful things a leader can do is to admit when they don't know something or the system itself needs fixing, as that honesty creates safety.

Excessive Workload and Lack of Resources

In helping professions like healthcare, education, or emergency services, many workers expect to face high demands. They accept that the work will be hard, involve exposure to traumatic events and at times be heartbreaking. What they don't expect is to be under-equipped, under-supported, and left to carry an impossible workload with little attempt by the organisation to fix these issues, other than lip service and a free massage or yoga lesson.

Several participants described moments when a lack of resources directly put them and others in harm's way. In one example, police responding to what they thought was an active shooter were issued faulty protective gear and substandard weaponry. They were willing to face danger, but they had not expected failure to provide the necessary equipment to perform safely.

More often, though, the damage was a slow burn. Overwork led to no rest time, no space to reflect on experiences, and no capacity to care in the way they imagined they would when they constructed their moral covenant about work. When staff are spread too thin to live up to their own standards, it becomes morally exhausting. It can be a cruel reality to be asked to care deeply while being provided with less and less. What is worse is that many felt this overwork was ignored or even normalised by management.

While frontline leaders sometimes made efforts to protect their teams, many didn't have the power to make fundamental changes. But participants agreed that even when the pressure couldn't be fixed, it made a difference to have someone in the leadership team acknowledge it, thank them, and show they understood the cost. Workplaces can't always eliminate stress. But they can recognise when it crosses into harm and take action about it. To do anything less is to create a workplace that is bound to injure its staff morally.

Poorly Applied Reward and Recognition

Most people don't enter a job just expecting gratitude, and many of the best technicians would say they don't do their role for the thanks. Yet, our need for meaning-making and self-esteem means we all need to feel seen and acknowledged. But when done poorly, recognition can feel like more betrayal. Many workers told stories of being honoured in ways that felt performative, tone-deaf, or deeply disconnected from their lived reality.

Nowhere was this sentiment expressed more deeply than with medals awarded for exceptional or long service. First responder organisations often award medals for bravery or long service; however, many participants reported that these gestures felt almost insulting. One received a long service medal through internal mail, without a word from a supervisor. He reflected that he thought the medal to be so perfunctory that he threw the medal into his locker, got into his work vehicle, and continued with work. Another individual was awarded a commendation for a job well done, despite having been denied equipment to make the job safer. He reflected, 'We didn't want recognition, [...] we wanted the right gear'.

Unsurprisingly, some I spoke to refused to accept the medals rather than go through the charade of getting something tokenistic. Others received their medals but were left with a burning moral pain. The problem wasn't the medal; it was more often the arbitrary criteria on which it was awarded, and that it represented a system preferring symbolic gestures over genuine gratitude. Arbitrary thanks are worse than no thanks; they have been a betrayal of what is right ever since Achilles had his valour stolen so publicly.

Recognition that is helpful is not a one-size-fits-all approach. Recognition that resonates with workers is relationally meaningful and personally respectful. Small but genuine acts, such as verbal thank-yous, personal notes, shared meals, and a helping hand, or quiet moments of acknowledgement, often go further than any formal award. When these simple covenantal acts are missing, it can feel like a betrayal. When formal recognition is given, it must be accompanied by proper support for those who may be grieving, angry, or traumatised by the events being honoured. Workers deserve to be treated as whole people, whose emotional labour is seen and valued. Recognition that truly helps must be based on covenantal, not contractual, principles.

Injustice, poor leadership, overwork, and hollow recognition cause more than stress; they are the elements perceived as betrayal that can wound people at a moral level. When our deeply held moral expectations are violated by the systems we serve, the result can be lasting harm. However, there is hope because the same systems that cause moral injury can be reshaped to prevent it. Leaders can change, cultures can shift, and recognition can become relationally real. In reality, betrayal and moral injury are unavoidable if we engage with others in a fractured world. What we can do is prevent it where we can, be aware of it, and understand how to heal best when it does occur, lest we avoid betrayal and fall headlong into distress, damnation, or death.

The Impact of Betrayal

When one's deeply held moral expectations are betrayed in the workplace, it is not merely a blip on our radar. Workplace betrayal has a profound impact on our identity and meaning-making structures. As mentioned in Chapter 2, the effect of betrayal on a person's identity has many negative impacts on their behaviour. The field of moral injury is more than a theory researched in academia or a therapy regimen for clinicians; it can be a profoundly personal existential crisis.

For those who have endured it, moral injury can leave scars that run deeper than many physical wounds, with some experts and survivors calling them dents on the soul. Being morally injured reshapes the way one sees oneself, their work, their colleagues, and the moral world around them. One author has described moral distress as a 'polite-sounding term for a gut-wrenching experience' (Berlinger, 2013). That phrase captures what many people struggle to express: that what happened to them at work did not just leave them stressed or tired, it left them completely altered in the mind, heart, and soul.

In my research, many participants described a deep inner conflict that affected their sense of identity. For some, the sense of detachment from themselves and others became so profound that they began to feel lost and numb. One first responder said, 'I went from being someone with an identity, literally overnight, to not knowing who I am, what my purpose is, and it is a constant internal struggle'. That loss of identity was not simply a crisis of confidence; it was a dismantling of the self brought about by an accumulation of experiences that felt deeply wrong.

This experience is not just emotional; it also has philosophical weight. The philosopher Paul Ricoeur once said that we are not the authors of our lives, only the narrators (Ricoeur, 1986). He helps us see that forces beyond our control often shape our life stories and are out of alignment with our moral covenants. This lack of control can be excruciating for people who have chosen professions built on responsibility, skill, and purpose. First responder organisations, for example, often appeal to recruits by offering a strong identity built on competence, bravery, and meaning. Recruits may come to believe they are beyond the weaknesses of others and are heroic. The realisation that they are not, especially when it comes through betrayal or systemic failure, can be devastating.

Moral injury often begins when institutions prioritise contractual obedience over moral consistency, or when command-and-control leadership overrides conscience, so that operational decisions demand moral compromises. This loss of agency is a profound psychological and spiritual injury that impacts our very moral character.

As far back as the 1990s, Shay reported that moral injury could damage a person's moral character and sense of identity, with others since expanding on this idea. One policing expert, Kevin Gilmartin, described how idealistic young officers, full of strong ethics and enthusiasm, can gradually shift into being cynical veterans who justify questionable behaviour as necessary or normal – a tacit act of vengeance against those they perceive have wounded them. This descent of character is not inevitable. An important self-protective factor in preserving one's core values is by grounding one's identity in life in family and community, faith, or other noble causes.

There can be serious legal and organisational impacts from acting out of vengeance at perceived betrayal. A danger for those who feel their moral covenants have been broken is that, out of a sense of anger and vengeance, binding contracts are broken. When this happens, we can self-sabotage and become liable to sanction. Vengeance taking actions include absenteeism, presenteeism, workplace deviance, poor organisational citizenship behaviour, and even criminal acts of violence. If this pattern continues, eventually one can become the betrayer of others by doing unto others what has been done unto them. Thus, the spiral of betrayal continues its downward trajectory as the betrayed becomes the betrayer.

In the stories I studied, the personal impact of moral injury was not decidedly concrete. Nine participants had received a diagnosis of post-traumatic stress disorder (PTSD), ten spoke of turning to alcohol or other substances, six attempted suicide, another three reported experiencing suicidal thoughts. The connection between moral injury and suicidality was very clear.

Workplace stress was not limited to the job itself. For many, it bled into their personal lives. Six participants spoke about how their relationships with partners had suffered, and some reported that break-ups pushed them even closer to the edge. Their experiences reflect more than personal sadness, capturing a common thread in moral injury – the breaking of connection to others, to one's own values, and to meaning itself.

The emotional effects of moral injury were powerful and varied. Sixteen participants reported feeling anger, frustration, guilt, shame, or emotional numbness. Others spoke of moments when their emotions boiled over. Many of these responses were explicitly linked to organisational failures. For instance, being unable to make decisions because of low rank caused frustration and resentment. Others described outrage at leadership decisions that put lives at risk. However, of all the impacts, guilt and shame were the most common. One told of feeling overwhelming guilt, shame, and regret, which led to a lack of sleep despite hating being awake.

These reflections show that moral injury is not simply occupational disappointment. Moral injury often shows in those who most deeply care but may not have examined their own moral framework about that caring attitude. Moral injury then affects how people understand themselves, how they relate to others, and how they find meaning in their lives. It can lead to broken relationships, self-destructive behaviour, and, tragically, sometimes even the loss of life. However, acknowledging these impacts is the first step towards healing. By giving voice to these stories and recognising the emotional truth behind the data, we begin to create a space for recovery.

Profile of a Betrayer

By now, you should have a better understanding of what betrayal is, what acts can be considered betrayal, and the impact it has on one's sense of self. Without exception, the betrayals that have been described so far have come from trusted people – loving partners, noble companies, leaders, and altruistic organisations. All the way back to Brutus and Judas, betrayers draw people close as a necessary part of betrayal. Part of the process is that we can become blinded to potential betrayal from those with whom we have affectionate regard. This irony brings us to a significant problem.

How can we trust anyone? More acutely, how do we live in this world without some constant neurosis that we are about to be betrayed by someone we love and trust? The short and unsatisfying answer is that we cannot live an everyday life without some fear of betrayal. At the start of Chapter 1, it was noted that C.S. Lewis said there is a way to protect your heart from betrayal: do not love anyone or anything. This is perhaps too high a price to pay to avoid betrayal.

We each have a choice: either we will be betrayed at some stage in our lives, or we will become so isolated and anxious that we will not be worth betraying. We only have this choice before the acids of betrayal start etching their way through the contours of our unexamined character. If we fail to understand what betrayal and recovery look like for us BEFORE we are betrayed, we may lose the ability to recover altogether. Preparation for betrayal requires wisdom, grace, and the willingness to explore ourselves at the depths of our deeply held moral expectations.

The choice to prevent lasting moral injuries requires the audacity to live a life of loving vulnerability with open eyes. It also involves making the vow not to hand over our moral code to the one who has betrayed us, that is, not to lose our sense of self and behave just like the loathsome creature who betrayed us. It's a rocky road at times, but the journey will be worth it. Having admitted the inevitability of betrayal, we need not simply surrender to fate. A small bit of wisdom can go a long way to preventing many of the betrayals we might experience in life.

Chapter 9 will take some time to detail tactics to defend against the degradation of our character; however, the final part of this chapter outlines the nature of noted betrayers and signs they are moving in for the kill. It answers the questions, 'who are the kinds of people we should be careful of giving our trust to?' and 'what will be the signs we should be wary of?'

But first, notice an important nuance in that first question. It was not a definitive 'who should I not trust'. Universal and unconsidered distrust is the first step on the road to an irredeemable heart. Instead, it is essential to consider ways in which to be judicious and how to strike with grace and conviction proactively.

Sociopaths and Psychopaths

Sociopathy and psychopathy are non-clinical ways of describing what is known as antisocial personality disorder. Both sociopathic and psychopathic tendencies can be identified mainly by their difficulty or inability to accept blame or show remorse. People with this trait will either deliberately or carelessly transgress social norms, then not care about it. These people will see their wrongdoing as either someone else's wrongdoing or, at the very least, someone else's problem.

It is important to note this trait of blame shifting while avoiding the temptation to provide amateur diagnoses about people's mental health and character. Therefore, be careful of labelling some *person* you know as a sociopath. Instead, take note of destructive *behaviours* rather than providing amateur assessments of psychopathology in individuals. With that caveat in mind, consider whether you've ever witnessed a family member, friend, colleague, or leader do something wrong and then fail to take responsibility for it. We all tend to avoid blame, but is the blame shifting moving towards outright lies and a settled way of acting? If so, watch out!

More telling than the occasional denial, of which we are all guilty, is when the denials of fault become a recurring pattern of behaviour. It is worth considering how someone takes any mild criticism. Finding this trait of blame shifting in people you know is probably relatively easy, but what about you? Are you the person who denies, shifts blame, and even tells a couple of white lies to avoid taking responsibility? The best place to start reform is with yourself. Taking blame in a punitive workplace has consequences, so begin this practice of admitting fault in safe relationships. I encourage you to take stock of your own reactions to criticism or blame from those who love you. Changing oneself is a point at which we truly can become the change we would like to see.

Blame-shifting can become a systemic corporate trait. Workplace cultures can become defensive of their public reputation and shift blame to others, as Part 2 will discuss in detail. Have a careful look at corporate ethics statements. Do any include 'curiosity at criticism', 'truthfulness', or 'taking responsibility for corporate wrong-doing'? The answer will be no for most workplaces. Therefore, look at how the PR department at your organisation works in a crisis. Is it 'deny, deny, deny' when wrong-doing is alleged until the weight of evidence is so strong that denial looks worse than admission of fault? As a rule, it is wise to give more trust to those who make public apologies than to those who make private excuses.

The Hurting

The hurting are people who have been betrayed or believe they have been betrayed before. At first glance, this is a strange group to suggest being cautious about giving trust to. Surely these people would want to spare that pain for others. In fact, they may betray others because they are displacing their anger from their betrayer to others, or have come to believe this is the way the game works now. They are dangerous because in their need, we draw near to help but end up becoming close enough to be betrayed.

We must tread carefully, as people who have been betrayed and abused need help, not continued rejection. However, this help must be given with a sense of wisdom as the hurting are a double-edged sword – those who have been betrayed and abused, especially during childhood, are more likely to keep being abused AND become abusers themselves. The World Health Organization reports that those who have suffered childhood abuse or neglect are often more likely to harm others in the next generation. Similarly,hose who are morally injured and who seek vengeance may turn their blowtorch onto you.

Ask yourself, who has been most overlooked for promotion in your organisation, or has just missed a promotion. Who in your workplace needs constant attention to their grievances? Who is the first person to show cynicism and dismissal of any work-place directives? Be cautious of those who attempt to draw you into their grievances and continually seek your time to complain. A gracious response to these people is necessary, as they have real hurt. Therefore, I recommend offering empathy, but be cautious about trusting them entirely or too quickly. If unattended and unhealed, these hurting people can become the third group of betrayers, the entitled.

The Entitled

The final group in this triad of potential betrayers are the entitled, who can be drawn from both anti-social behavioural disorders and those who have been hurt. Both groups can feel they are owed something and are entitled to receive their dues in any way possible. A way to spot these people is often by the use of their collective pronouns – 'if *they* [...] then *I*' followed by negative blame shifting talk – 'If *they* hadn't refused my overtime, then *I* wouldn't be taking sick days to play golf!'

Instead, people who use 'we/us' with a curious tone with those both up and down the work hierarchy are more reliable than those who use 'them/they' with cynicism and anger. For example, 'we are all going to solve this together'. These are people who curiously and inclusively ('us/we') still consider themselves part of a unified community committed to common goals.

Those who use 'they' a lot may also use 'us/we'; however, they do so in the process of leveraging your loyalty to cajole you into their pact of discontent. It might sound like, 'we aren't going to take this from them, are we'. Note that there is no question mark of curiosity; it is a kind of undeniable demand that you join their faction. Those who descend into 'theyism' become disconnected from those they believe have wronged them and will seek your loyalty until it is no longer beneficial. Tread wisely here, *you* do not want to become their *us* nor become their hated *they*.

Layson's Triad of Caution

Having listed three noted categories of betrayers, it is worth noting my own personal list of those who betray: sycophants, show-offs, and bullies – 'Layson's triad of caution'. Sycophants are individuals who consistently seek to curry favour with those in positions of power or influence, often showing a disregard for those they must overstep to achieve their goals. Show-offs crave popularity at almost any cost and they will drop you like a gun if they think you might become a liability to their popularity. Bullies are people who get up by pushing people down physically, psychologically, spiritually, or emotionally. Watch out for them. I believe these personality failings exist in a Venn diagram that reveals even more nuanced categories of workplace betrayers. Figure 3.1 shows how these three traits intersect and the niche categories of betrayers that exist in the liminal spaces.

The bully/show-off crossover is the person who will take your glory for themselves. You may have done group assignments with them at college or worked in a team with them for some time. They are recognisable because they are 'I specialists', talking of almost nothing but themselves and the achievements they claim in the first-person singular. I specialists may not blame others, but they will proactively steal your ideas and any credit that is due to you.

The show-off/sycophant cross-over distorts the we/us usage to create division – a different kind of 'we specialist'. They will drop the name of some influential figure they were in a room with once, as though the two of them are 'we'. These people will love to talk about experiences and beliefs they have in common with those who hold real power and may even drop your name if it benefits them. Be wary of these people, particularly if they are lower than you in the organisational hierarchy, partly because they want your job.

The sycophant/bully cross-over is the boss's fixer of broken things. They are not always natural bullies, but in a desperate desire to get proximity to power, they are useful idiots for the real bullies who devise ways to avoid blame. If the fixer is cajoled into an act of betrayal, the instigator can deny any responsibility. In some ways, I feel

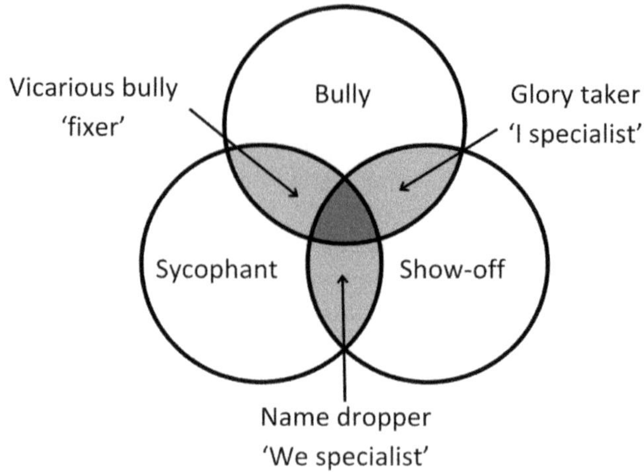

FIGURE 3.1 The Layson triad of caution.

great sympathy for fixers, as they often become scapegoats. Either way, watch your back when around the boss's fixer before they fix you.

Signs Your Betrayer Is Readying to Strike

As we conclude this chapter, we take one more step to help us recognise when people are about to pull out the knife of betrayal. There are a few patterns of behaviour that wisdom tells us we should be wary of. But first, remember betrayers are those you trust and even love. Betrayers often act in ways that do not seem harmful at all but seem highly friendly and helpful. The irony of betrayal is that the friendly and helpful are the ones who betray most dangerously – these personal kindnesses are known in some fields as grooming.

Grooming behaviours are trust-building activities that can include special treatment and praise. Groomers will help you get opportunities and experiences that you would not have without them. Groomers act in this way because they want you to say, 'Oh, they are such a nice person, they would never…' when accusations about them arise. Chapter 7 will make note of the importance of grooming in the process of scapegoating, but for now, it is worth noting that the first act of a betrayer is to increase trust. They will use generosity and flattery to gain your trust.

Flattery is particularly deceptive to us because it builds our own ego and can be recognised by a simple test. Beforehand, it is worth noting that flattery is always done in front of you so you can hear the praise and be puffed up by it. But flattery has a flip-side – gossip. I have heard it said that flattery is saying something in front of someone's face you would never say behind their back, while gossip is saying things behind someone's back you would never say to their face. Therefore, to work out if you are being flattered take note of what that person says about people who are *not* in the room. If their language is regularly derogatory of your friends and colleagues not in the room but generous about you, then the odds are that you are being flattered as part of the grooming process. What is more, their gossip about you may resemble what you are hearing about others.

There are some genuinely kind people who will look after your welfare and will encourage you to your face and so we want to avoid being hyper-vigilant about every act of kindness done for us. Therefore, to detect betrayal at its early stages, it is vital to recognise what happens after the trust-building phase. Here three activities are key to notice: boundary testing, secrecy, and demands for loyalty or in-group compliance.

Boundary testing involves saying or doing things that are a bit morally risky. Testing a boundary might include physical proximity with you that is a little uncomfortable. Boundary testing occurs when a pat on the back becomes a lingering semi-massage, or when you are encouraged to participate in a small act of workplace deviance – when good favour starts to be leveraged so you become compromised in some way. Like teenagers being tempted to cut school with the cool kids, these boundary testings make you feel included whilst also complicit.

The second action that feeds out of this boundary testing is secrecy. Having a secret with someone can be enthralling because we feel included in an elite inner circle. However, secrecy is a two-edged sword. Researcher on shame, Brene Brown, notes wryly that 'If you put shame in a Petri dish, it needs three things to increase exponentially, secrecy, silence, and judgement' (Brown, 2012). While our ego is being scratched by knowledge of secret information, the ground is being laid for shame and possibly betrayal.

The final part of this process is the call for personal loyalty. Remember, loyalty is not the opposite of betrayal, but it can be its precursor. The lesson here is: do not be sucker punched into personal loyalty to a psychopathological character. In place of interpersonal loyalty, seek covenantal faithfulness as described in Chapter 2. Instead, commit to a higher moral calling that is independent of you both, such as truth telling. Covenantal faithfulness avoids the pitfalls of personal loyalty and calls all parties to aspirational moral virtues.

ITEMS FOR CONSIDERATION

- Take the next week to consider the 'they', 'us', 'we' language of your senior leaders. What do you think it says about their trustworthiness?
- Who was the last person to ask for or demand your loyalty? Are they more mafia boss or trustworthy ally, and how can you tell the difference between those two?
- What are the kinds of moral virtues that you aspire to that could become the independent part of a relationship based on covenantal faithfulness?

CONCLUSION TO PART 1

In Part 1, we have discovered why betrayal, particularly workplace betrayal, is so corrosive to our sense of meaning and identity. The founding conviction of this section is that it is very hard to change the world, but relatively easy to change ourselves. As you have answered the questions at the end of each chapter, it is hoped that you have started to develop a robust framework to examine and guard your moral self and sense of meaning. As we move into a more pointed exploration of workplace betrayal and its dynamics, you may find you need to keep referring to this moral code you have started to explore.

Part 2

The Mechanisms of Organisational Betrayal

4 Faking Good
The Halo Effect

A true word

> One of the oddest aspects of organisations is that they often behave in ways that are precisely the reverse of their intentions in the world. This is known as their 'shadow', and it often shows up as the awful and shoddy treatment of workers. It is the dark side of good intentions. And it's often the 'best' organisations that have the deepest shadows.
>
> *Fuller, 2009*

INTRODUCTION: THE DEATH OF INNOCENCE

In March 2007, 17-year-old Takashi Saito began his dream career as a Sumo Wrestler in one of Japan's most prestigious training academies. For a young man of his large physical build, the move seemed appropriate and honourable. Besides the highly physical nature of Sumo Wrestling, Sumo carries with it an air of holiness and national pride. The spiritual and national significance arises from Sumo's Shinto heritage that still permeates every element of the fighting. The ring where bouts are fought (dohyo) is elevated to the stature of holy ground with paraphernalia reminiscent of a religious shrine. At the start of a fight, the ring is blessed by a priest to demonstrate that Sumo is more than just a fight to bet on – it is a sacred event.

To become a wrestler, Takashi underwent strict training, with the possibility of fame and the honour of being a revered champion wrestler. By June 2007, three months after commencing, Takashi was dead. He died from injuries sustained as part of punishment for absconding from the harsh treatment he was receiving. The push to reveal what happened was strongly resisted, with many considering the training academies to be beyond questioning – until a father's tears led to a national outcry.

It was finally revealed that on the evening of 26 June 2007, Takashi had been beaten to death with beer bottles and a baseball bat. The assaults, humiliation, and murder were hidden behind the veneer of holiness. Takashi must have believed he was entering a world of honour, discipline, and holiness, but the truth killed him. Since Takashi's death, this veneer of honour has been shown to cover over a culture that is rife with violence, corruption, and betrayal.

For fear of betrayal by corporate monsters, many regular folks flee from organisations or movements that have poor public reputations. For good reason, we seek organisations

DOI: 10.1201/9781003499725-6

we think we can trust, those we see as having a halo of sincerity, purity, and even sanctity. But as Takashi Saito tragically discovered, a good reputation can be lethal for those who fall under its spell. As many who have been betrayed at work discover, perceptions of purity and morality are often used as cover to hide ugly truths.

The perception of purity in organisations is a form of what some call the 'halo effect'. Here, a good reputation for one attribute, such as a person's height, tends to increase perceptions of other unrelated elements, such as intelligence. This halo effect was famously recognised and labelled by behavioural psychologist Edward Thorndike (Thorndike, 1920), who noticed a 'constant error' in earlier psychological ratings in a paper where superiors rated the intelligence, leadership qualities, physique, character, and so on, of employees.

Thorndike showed alarmingly high correlations between completely unrelated aspects, such as intelligence and physique, writing that 'a halo of general merit' was transferred from one trait to other unrelated traits.

The practical outworking of the halo effect is often reported in research. For example, taller people are more likely to be promoted and paid better. Such transference of attributes can combine negatively with our inner prejudices. Research over the last 20 years consistently shows that job applicants with African American–sounding names, such as Lakisha or Jamal, are 2%–50% less likely to get interviews than applicants with white-sounding names, such as Emily or Greg (Bertrand & Mullainathan, 2003; Kline et al., 2022).

This bias confirmation illustrates how the constant error can have both positive and negative effects. A negative perception of one aspect can lead to a comprehensive evaluation of an employee being negatively impacted. However, our positive associations can create an organisational halo effect that is often a precondition for the worst betrayals in the workplace. When we start working for an organisation that does 'good' and assume it must be a good workplace, we can be set for a significant fall.

Halos

Halos have an ancient provenance, having been part of religious and mythical iconography throughout the ages, across many cultures. Whether it be an ancient Greek god, a Hindu deity, or a Christian saint, artworks have depicted halos of light surrounding revered characters. Halos serve as a visual cue, allowing viewers to quickly discern which characters are worthy of our awe and even worship. Halos convey a simple message: 'You can... *should* trust me'. Halos are highly powerful ancient memes that portray a good reputation, but are, in reality, fictitious, fanciful, and ultimately unreliable figments of our moral expectations. The more we buy into halos, the more susceptible we are to betrayal.

While religious and artistic depictions of halos appear to be less common, the use of the word halo in literature is at an almost historic high.[1] It is ironic that in a supposedly rational and scientific age, we speak more regularly of mythical memetic halos. One explanation for this is that modern organisations regularly create, curate, and communicate halos of faithfulness around their mission and ethos.

Organisational halos are not depicted in artwork but disseminated through spin. The creation and maintenance of halos allow organisations to manage their brand and employ public relations experts to curate a benevolent public image. Halos are also the first ingredient in creating workplace betrayal – remember, only someone you trust can betray you. Organisational halos engender and maintain trust to the casual observer. They can also be used to bury dissent or criticism, and more dangerously, cover over hypocrisy, abuse, and murder. Experiences like those of Takashi show we ought to be wary of organisations with the brightest halos – as they cast dark shadows.

Organisations cherish their halos, but they each achieve their existence through diverse means. Some of us *assume* that particular organisations are trustworthy simply because their altruistic mission suggests moral behaviour. Other organisations *manufacture* theirs as they know they will never acquire one naturally. Finally, a smaller class of organisations *earn* that halo by genuinely benevolent behaviour.

Assumed and Legacy Halos

In one territory of Australia, the catchphrase, 'Join the most trusted profession', used to be emblazoned on their ambulances. Surveys of the public regularly show paramedics to be the most trusted profession, which is the basis for their organisational halo. This ambulance service co-opted the halo effect for its recruitment strategy. It rang out loud and clear that this organisation was trustworthy. However, the public who awarded this status of trust gave it to the paramedics, not to this organisation. Perhaps a case of stolen honour.

The corporate entity adopted a positive attribute belonging to those who worked for it. However, not all staff of that ambulance organisation were willing to tow the corporate line about organisational virtue. In union-inspired defiance, paramedics emblazoned ambulances with unofficial slogans that said, 'The most respected, but least protected'.

Many paramedics felt betrayed by their organisation due to a shortfall in resources, staffing, and pay. To the organisation's credit, many of those issues have found a favourable resolution, but only after the expression of outrage and lengthy industrial action. First responder agencies and many non-governmental agencies that engage in philanthropic work are often assumed to have organisational halos. Recent research reported that the 'perceived sense of moral goodness within NGOs can lead to the NGO being glorified by its staff and volunteers, creating the belief… that it is morally better than it actually is' (De Bruin Cardoso et al., 2023, p. 2).

Joining a philanthropic or altruistic organisation contains a heightened risk of moral injury. Therefore, before embarking on a dream career with a 'trustworthy' organisation, there are important questions to ask yourself: 'Are my expectations of trustworthiness reasonable?', 'Is this organisation selling something about itself that it is unlikely to supply?', 'What would happen to my sense of self if this organisation is not as good as it makes out to be?' If the answer to any of these is a negative, then proceed with caution.

Manufactured Halos

Some industries, such as the aviation industry, are unlikely to be presumed righteous for the work they do – they often generate large profits, often at the expense of poor service. Furthermore, aviation contributes significantly to air pollution. Hanging a halo on profit, poor service, and pollution is hard work but one can be manufactured. Australia's national airline is Qantas, with whom I regularly fly and hope to do so in the future. It projects itself as a good organisation, but at times, the Australian public has trouble trusting Qantas. Qantas, therefore, works hard to manufacture its halo.

Since Dustin Hoffman's character in the movie Rain Man said he would only fly Qantas because it had never crashed, we have believed Qantas to be a safe airline. Indeed, Qantas ranked as the second safest airline in the world in 2025. However, safety is not enough to project virtue, so, like many of its corporate contemporaries, Qantas works hard to build a strong public reputation for its social values. To achieve this, Qantas highlights its role in fostering social connections by uniting families and friends, alongside its involvement in public social policy. Qantas was a prominent corporate supporter of marriage equality in Australia and later emblazoned its planes to influence a national referendum on rights for First Nations people. Qantas positioned itself as a virtuous and safe airline that could be trusted with issues like inclusion, sustainability, and diversity, and as a great place to work and do business.

However, in the shadow cast by this halo, not all was rosy. Courts found that during the early days of the COVID-19 pandemic, Qantas illegally stood down a workplace health and safety representative after he raised safety concerns for cleaners of planes. It appears Qantas was safe for passengers, but less so for its lowest-paid employees. The High Court of Australia also found that during COVID-19, Qantas illegally outsourced 1,700 ground crew jobs to circumvent union involvement in pay and conditions deals. And then Qantas was found to have misled customers by advertising flights it had cancelled before selling tickets. Perhaps the final straw, Qantas continued to pay multi-million-dollar bonuses to senior executives despite record cancellations, consumer dissatisfaction, and staff cuts.

The halo created had unbearable weight, which ultimately Qantas was unable to maintain. The juxtaposition of virtuous social campaigns and troubling corporate behaviour suggests Qantas may have embraced social causes as a form of reputational insurance, believing public goodwill could buffer criticism in other domains – that is, their halo would make people look the other way.

Qantas is by no means isolated in its tactics. Industries with poor reputations will weaponise good actions to cover over profits and self-protective instincts. Miners, corporate banks, first responder agencies, and the military often demonstrate their virtuous social values but then act in direct opposition to their professed liberal ethics. If you're considering a corporation that promotes diversity, equity, and inclusion (DEI) policies or strong environmental, social, and governance (ESG) credentials, don't just assume it is a safe place to work. The values underpinning DEI and ESG are important, but the weaponisation of these noble aspirations serves to create a culture that is ripe for betrayal.

Earned Halos

Not everyone is out to get you. Yes, many bad workplaces manufacture halos to obscure their broken moral compass. Often, many good people work in these troubling organisations and are battling the weight of moral dissonance. Other organisations have rightly earned their halos through moral and relational clarity, and they are the places it is a pleasure to work on most days. The delight rarely comes from the work itself, nor the supposedly virtuous organisation, but from good colleagues and leaders. Good workplaces are a factor of the people you work with, as people have halos, not the company.

Your colleagues and leaders situate any workplace in the bearable to beautiful range for two main reasons. Firstly, the organisation you work for is simply shorthand for the culture and outcomes of the people working there. Secondly, as noted in Chapter 1, humans are fundamentally workers who work best in social networks. Working in complete isolation, even for the most introverted person, poses a significant risk to our sense of place in the world and our ability to find meaning and connection. Saying a good workplace is a function of the people who work there needs further investigation, to understand precisely what it is about people that makes it pleasant.

The overriding element for a noble workplace is trust – the primary prerequisite to betrayal. Without love or trust, there is no betrayal. However, without trust, we experience no love or intimacy, so we must dig deeper again. Research continually shows that civility is a key indicator of who is safe to trust. Some may object, thinking that simply being nice is a weak indicator of trustworthiness. However, civility is a behaviour that many of us in an increasingly polemical world are understanding less.

Civility to people we agree with is flattery. However, loving those who love you is not real civility. Real civility occurs when people disagree with you, even firmly, but can keep a civility that neither curses nor excludes. In the workplace, this is seen when people disagree and then work towards a resolution and improvement. When a disagreement arises, the old football saying, 'play the ball and not the person', is applicable. That is, civil people will seek to address the issue, rather than descend into character assassination of the person.

When workmates are willing to hear a position that you hold, which is diametrically opposed to theirs, and then attempt to understand, we are starting to tap into civility. Civility demands that they not only understand you but could also explain your position better than you can. Reaching the point of civil disagreement rather than civil war requires empathy, active listening, and an overarching culture of psychological safety (see Part 3).

Trust is essential to have with co-workers, and organisations that truly earn their halos are those with trustworthy leaders. There are numerous traits we might expect in good leaders, including civility. However, three halo-worthy traits are most important in leaders: ability, benevolence, and integrity, with benevolence holding the key to trustworthiness.

Ability involves your leader's technical skills, and integrity consists of using those skills diligently and ethically. However, if a leader is not benevolent to their workers, ability and integrity mean very little. Benevolence is measured by the perception of whether a leader is concerned with their workers' interests as even more important than their own (Lleó de Nalda et al., 2016). Benevolence by leaders to their staff has been measured as a factor of several traits, including:

- the perception that one's leader is really concerned for our personal fulfilment,
- that the leader desires to meet our needs in the workplace,

- a leader genuinely wanting to know what is important to us,
- that a leader would never knowingly do something that would damage our interests,
- and finally, that they would leave their own work to help us (Mayer & Gavin, 2005).

Having now examined the varieties of halos that organisations acquire and manage, we must heed a critical warning. No matter the method by which a halo is acquired, all such organisations have taken the first and pivotal step towards workplace betrayal, because it only happens with those we love and trust. To truly understand how halos are abused, we must examine the entity itself, rather than just studying the halo. Organisational culture and morality are influenced by the nature of organisations, their leadership, and their level of maturity.

ORGANISATIONS: THEIR ESSENCE, LEADERSHIP, AND MATURITY

Do Organisations Even Exist?

Let's start this section by making a counterintuitive statement: organisations don't *actually* exist. Much like halos, the organisations we work for are a figment, a concept, an idea. We can see the buildings where workers congregate, buy shares, and purchase their products and services. However, organisations are transitory and inanimate. If you do not believe this, try to find the place where the last remains of Enron or Lehman Brothers are buried. Organisations are not sentient beings, nor carbon-based objects. Despite this, it is common to attribute human actions, such as leading, designing, or even betraying, to organisations. This transference of attributes to inanimate objects is called anthropomorphism.

While it is helpful to recognise what organisations are not, it is more helpful to understand what they are. Organisations are merely a manifestation of the collective hopes and aspirations of a group of people with a like-minded approach to a task or mission. The relationship between people and organisations is dynamic, with each influencing the other for better or worse. At the most basic level, organisations are a representation of the people who work in them, and those with higher ranks or positions have more influence on organisational culture.

Therefore, anthropomorphisms that suggest organisations perform actions, either good or bad, can ultimately be used to hide the fact that organisations don't actually do anything – rather, people perform actions and create culture. The use of anthropomorphisms for organisational culture can occur to cover over wrongdoing. Blaming an organisation for a bad culture is the dark side of organisational non-existence that is used to avoid personal accountability. Those high enough in the hierarchy may blame organisational culture as a technique to prevent themselves from being sanctioned, and do this with great effect.

An example of this blame-shifting is seen in the findings of an Australian Royal Commission into the banking industry, which found a widespread culture identified by the pursuit of short-term profit at the expense of basic standards of honesty – at its worst, it failed to prevent money laundering alongside triggering concern about banking institutions breaching counter-terrorism law. These are serious allegations that require accountability from people, but much of the blame was taken by the organisation and the pain was felt by investors rather than people.

Of the four largest banks in Australia, three CEOs resigned, some with million-dollar-plus exit payments. However, none were charged criminally or civilly with any offence. The organisation did not get off so lightly, and investors and customers were fined heavily. However, the culture was not organisational; it was human-created, curated, and secreted by certain leaders until a Royal Commission exposed it.

In a smaller way, I began to recognise that corporate culture reflects those who hold power within organisations. While I was a firefighter, I also secured part-time work tending to indoor office plants to earn some extra cash. Although this wasn't particularly exciting, I gained access to almost every office within an organisation's headquarters. I went to the mailroom for the little plants, to the foyer's display of impressive palms and to the CEO's corner office for exotic displays.

Each workplace had a particular vibe, wherein I would be greeted in eerily consistent ways. Some offices I knew would be filled with staff who were dismissive of my presence. I would be granted permission to tend to plants with a wave of the hand and a grunt. In other offices, I was greeted heartily and engaged in conversation and civility. There were some exceptions to this uniformity, but what I did notice was that the general mood of any organisation always reflected the way the senior leader treated me – no exceptions.

While I maintain that organisations do not have being, they do exist in a legal and cultural sense. Thus, it is irresistible and even foolish to avoid talking about organisations. However, when I discuss organisations, I will be referring to the collective beliefs, aspirations, and actions of a group of people, particularly its senior leaders.

Although there are many factors contributing to the nature of any organisation, I wish to examine just two. Firstly, the dynamic I noticed in those plant-bearing offices was the key role of leadership. Leaders drive the culture of an organisation in a way that is disproportionate to their percentage within that organisation, largely creating the culture where betrayal is rife or rare.

Secondly, it is helpful to explore the evolution of organisations from start-up to legacy phase. Throughout the stages of our lives, from childhood to teenage years, adulthood, and old age, we remain the same person but change in the way we express those attributes. For example, the tantrum of a young boy, angry outbursts of a teenager, and sullen grumpiness of an older man mirror the stage of life that person is in. Likewise, organisations remain the same collection of core attributes but change in the way they express those attributes over time. This process means that the mechanisms and language of betrayal manifest differently at each stage of their evolution.

Leaders and the Organisation

In 1997, two significant organisations merged, resulting in a culture created by senior leaders of the new collective that would, 21 years later, contribute to the deaths of 346 people. In 1997, Boeing merged with its competitor, McDonnell Douglas, to form a larger, more competitive company that grew into an engineering behemoth. In

2018 and 2019, despite their proud heritage and engineering might, two of Boeing's newest jets (737MAX series) plummeted to earth, killing all on board. Was this the organisation's fault, or a result of human factors? Leaving that question open for now, let's examine some of the decisions of senior leaders.

As the new organisation evolved, its primary focus shifted, and one might say its culture transitioned from engineering excellence to profit-making. When the profit cart was put in front of the engineering horse, trouble was bound to follow. The primary symbolic change was the relocation of Boeing's headquarters from Seattle to Chicago, made to establish a new cultural direction. Notably, senior leaders were separated from the company's engineering work.

But who or what made this change and what was its effect? Did the culture change the location? Or did the leadership change the culture by changing the location? The question was answered most clearly by the Chief Operating Officer (COO) of the day: 'When people say I changed the culture of Boeing, that was the intent, so it's run like a business rather than a great engineering firm. It is a great engineering firm, but people invest in a company because they want to make money' (Defazio & Larsen, 2020). With the benefit of hindsight, these words about de-emphasising engineering excellence have been shown by multiple investigators to be unwise. Senior leaders changed the culture to one of profit, which contributed to hundreds of deaths. Culture is not created out of the ether, but by people, usually senior leaders. We will see in Chapter 7 who or what was held accountable for these sorry events.

If the organisation you work in has a bad culture, we must be clear that it is the product of decisions made by senior leadership. Culture is not some spontaneously arising vibe that cannot be controlled; it is a leadership issue. We should therefore select and enable leaders to make wise decisions about the culture they will create in your organisation. I point this out, not to blame leadership per se, but to recognise the significance of leaders in your work experience. Let's not view leaders as the problem but as the solution to workplace betrayal and moral injury! Sadly, many leadership roles are awarded to individuals with high technical ability, rather than benevolence, and are rarely based on their potential to have a positive moral and relational impact. As a result, many senior leaders are not trusted.

Inspired by surging protests against globalism, in 2000, Richard Edelman began measuring trust in society's foundational institutions: government, business, non-governmental organisations (NGOs), and the media. The Edelman Trust Barometer has been produced annually, revealing trends that are now being reported. In 2025, the barometer reported that a pervasive lack of trust, particularly in democratic countries, has led to the development of a grievance mindset among workers and citizens.

This grievance for perceived wrongs accelerates the erosion of trust in leadership. Only 30% of those with the highest levels of grievance believe CEOs can be trusted. Higher grievance also leads to polarisation and a resultant loss of civility to those with whom we disagree. Edelman reports that 40% of highly aggrieved people approve of hostile activism that includes spreading untruths, damaging property and physical violence. The 2025 report states that the social contract that maintains peace is at risk because leaders are not taking action.

The grievance reported by Edelman closely aligns with Jonathan Shay's observations of the morally injured, who felt betrayed by their leaders and the culture. Those who feel betrayed start to act out because they have lost social trust. What Edelman and Shay recognise is that leadership matters more than anything else in the corporate entities that shape our society. An increasing number of people are feeling betrayed, morally injured, aggrieved, and even entitled due to the wrongs committed by society's leaders.

It may be easy to blame society; however, betrayal is always of human origin. Organisations do not betray people – it is bad leaders who perpetrate and enable betrayal. Our world needs to pivot to one where this behaviour is not accepted but changed and improved. The truth is that leaders shape culture, and culture drives performance or grievance. Few leaders are currently enabled to be benevolent leaders who create cultures that don't harm people. Chapter 10 will examine good leadership, but we must now explore the stages of organisational development and its role in shaping betrayal and moral injury.

Organisational Evolution

Several times in my career, I have moved into roles immediately after the founder of the organisation or the first generation of staff have moved on. The first time this happened, I was warned about Founder's Syndrome, most especially the difficulty of following the founders of organisations formed for worthy causes. Founders possess the chutzpah and vision to achieve great things, as well as the specific skills and attributes necessary to get a job done. However, those who come afterwards change things – they alter processes, goals, and culture.

When a new leader takes over, they are faced with several pressures regarding what they have inherited. Like no one else, a succeeding leader sees the dirty laundry left by their predecessor. A leader following a founder often finds a lack of due process and governance, for which visionary genius may have little desire or skill. Third-generation leaders often find a diamond in the rough, characterised by good goals and processes, yet they lack the elements necessary for sustained growth. Fourth-generation leaders don't want to break what they have inherited. Each successive General Manager or CEO has different pressures.

While leaders make the single most significant impact on the way organisations function, they can be constrained by what they have been handed. I have come to realise, from working in second-generation leadership roles as well as in legacy institutions, that workplace betrayal manifests in different ways at various stages of an organisation's lifecycle. We will examine four stages of an organisation. These four phases were first brought to my attention in a 2009 speech by the controversial founder and preacher Mark Driscoll.

Driscoll planted what became one of the biggest churches in the world. He visited Australia on a speaking tour at the height of his powers, before many questionable elements of his leadership style were revealed. Strong allegations of deplorable behaviour by Driscoll have never faced court, but did lead to the disintegration of the church.

However, Driscoll is well-read, intelligent, and possesses deep insights, with considerable chutzpah. He is a founder by nature. He noted that organisations tend to start as movements, become organisations, then proceed as institutions, and finally become *museums*. From memory, Driscoll did not reference anyone in particular about this model. However, he may have been channelling people like Howard Snyder and Vance Havner, who described organisational evolution as a visionary *man* who becomes a *movement*, a *machine*, and eventually a *monument*. We will utilise Driscoll's four phases to illustrate how betrayal manifests in an organisational lifecycle.

Movements

For 10–15 years, an organisation could be considered a movement depending on two factors: firstly, how rapidly the idea and product become successful and financially viable, and secondly, how long the founder remains in the leading role. If you have joined a movement, it was probably not for job security, but rather out of deep conviction or faith in the higher ideal being pursued. Moral convictions often drive movements in which those who join have an almost existential stake.

The mission of a movement is often driven by leaders of great vision and personality who can attract willing followers who pour themselves into the common cause. Early on, the boundaries between personal identity and organisational mission often blur. Sacrifices of family, friends, and health are all worn as badges of honour, and loyalty to the vision is core. People bring their deeply held moral expectations to a way of being, not to the organisation, which is so new and so disorganised.

But the intimacy, conviction, and shared vision can be weaponised. Founders can demand unquestioning allegiance to themselves and to the cause because distinguishing between the two is challenging. Any questioning of priorities or methods can be dismissed as disloyalty to the leader. Leaders can justify poor treatment of staff or even immoral and illegal means for the sake of the common visionary ends. The phase is very organic and often chaotic, with little time or inclination for due process or even safety. There is a high appetite for risk, or at least a low appetite for any process that might restrain the dream.

At this stage of organisational growth, betrayal manifests as a betrayal of the faith we have placed in the lofty morals that underpin our shared goal. The ideals, promises, and relationships that initially inspired commitment hold little weight when people push back against the charismatic leader. The German theologian Dietrich Bonhoeffer, martyred by Hitler in the last days of World War II, abhorred this thinking in both his native Germany and in his beloved church. He wrote the following about founders and visionaries of nations and churches who are consumed by a dream:

> Those who love their dream of a [...] community more than they love the [...] community itself become destroyers of that [...] community even though their personal intentions may be ever so honest, earnest and sacrificial. God hates this wishful dreaming because it makes the dreamer proud and pretentious. Those who dream of this idolised community demand that it be fulfilled by God, by others and by themselves.
>
> *Bonhoeffer, 1954*

In this phase of an organisation, people are highly susceptible to betrayal, and hopes and faith are used against them with phrases like 'we've worked so hard for this vision, and it is happening… surely you would not want to risk our hard work to make a big deal about this issue'. Those betrayed may have felt that their moral compass was compromised by the means used to achieve what they believed was a morally good end.

It can also feel like the movement lurches from one crisis to the next. In fact, creating a crisis is a tried-and-true technique by cunning leaders. Crises require leaders to take tighter control of all aspects, which can hinder genuine openness and accountability. These politically savvy leaders might say, 'once we get through this situation, then we can address your concern', or 'once we get over this hump, we can get more resources'.

While the movement is still in its early stages, the danger of betrayal and moral injury is high because moral expectations are high. It is worth reflecting on what you are being asked to sacrifice for someone else's wish dream, and whether the dents in your soul are worth it. If you and the movement you love survive this early phase and the founder's wish, and become an organisation, the dangers of betrayal take on a different form.

Organisations

This shift to organisation typically occurs when market share is generating profits, growth is steady, and the founder either transitions out or establishes a board of directors to oversee operations. The hectic mentality of the movement phase is not sustainable now that the mission has become accepted and profitable. The organisation gets either new middle management or new leadership. A staff team is hired, policies are established, and systems to manage the ever-increasing complexity are implemented. During this transition from visionary to strategic, the mood also changes.

Without a change to strategy and process, it is unlikely that positive early work will survive. Business coach and author Michael Berber writes that there is a shift from entrepreneurial spirit to a need for better management. While this is often necessary for survival, it introduces tensions and creates new opportunities for employees to experience moral injury.

Betrayal is now perhaps more subtle, but still carries the possibility for moral injury. Founding members may watch in horror as the purity of the original mission is diluted in the name of sustainability. For those present during the movement phase, there is a feeling that efficiency is replacing the purity of the original vision, and process is replacing relationship. Newcomers may openly question how good things could have happened so haphazardly or how luck outweighed good management. They will notice the lack of attention to physical and psychological safety and start to hear stories of people pushed down and even betrayed by the founder.

Freedom now comes to those previously silenced. For the first time they can voice their concerns in the open to fresh ears who value good management as much as visionary leadership. At this stage, old values and new systems, as well as old and new staff, are competing. I have seen this conflict many times, and it can feel like a battle for moral high ground.

For founding members, moral injury arises from a perceived betrayal of purpose. When the organisation justifies these shifts as a sign of maturity, it can feel like a betrayal of everything it once stood for. For newcomers, the betrayal can arise from the constant battle between old hands and old cultures who resist accountability and transparency as a lesser good. It takes a wise and patient leader to navigate these factional concerns. Chapter 9 explores how enlarged thinking can prevent moral injury and mitigate ideological warfare during this phase. In any case, if your organisation is fortunate enough to survive, it may become an institution with different sets of values and opportunities for betrayal.

Institutions

Eventually, a successful and growing organisation becomes an institution – such as government agencies, large banks, the military, first responder organisations, or health providers. In this phase, battles to produce due process and governance are largely over, and the organisation has become recognised and stable, with some level of brand recognition. It now has a department for every function, initially run by a single person. There is a dedicated safety team, finance department, and human resources team to develop glossy ethics statements.

Some people working in these departments are aware of the institution's core business, but not all are. These departments are primarily separated from operations and siloed from one another. Several visions of good replace the unified vision of the movement and organisation. Each department battles for money and glory, knowing that every cent their team gets is a cent that others do not. Here, the first possible source of betrayal emerges: internal factions and interdepartmental politics.

Not long ago, I became involved in research in a military context. I came in naively, assuming the military created the concepts of unit cohesion and esprit de corps. After a couple of years, I saw that militaries rarely need foreign enemies; they are busy fighting each other. The Navy despises the Army, the Air Force patronises the Navy, and the Army marches over both. Any internal dissenters are converted to enemies in the mind of the establishment and are vanquished through their extensive resources.

I wish I were joking, but Royal Commissions into banking, the military, and churches reveal how large institutions exploit their power to harm and suppress. Ironically, this ability to suppress is a function of the processes created to prevent harm. Institutions reach a point where they have developed so many processes, procedures, and policies that no employee can fully comply with them. If one speaks out, administrative violence that uses due process to crush dissent can be expected.

During the movement phase, there is only lots to gain. However, institutions have money, power, and prestige, so they have a lot to lose. Institutions, therefore, develop a powerful instinct for self-preservation through attacking even those who have a legitimate cause for concern. Institutions employ public relations experts to protect their reputation, legal departments to defend their systems, and middle managers to preserve the status quo. Institutions ultimately harm the very people they claim to serve.

During the institution phase, betrayal becomes systemic. Ethical concerns are redirected through bureaucratic channels that prioritise containment over accountability. Survivors of harm may be silenced, and whistleblowers may be punished or crushed by the process. Procedures and contracts may be followed diligently, while covenantal concerns, such as mercy and improvement, are distant concepts.

Due to the overwhelming weight of organisational pressure, moral injury manifests as disillusionment and a sense of dehumanisation. Staff become expendable, valued less than the organisation's reputation. Canadian researcher Lorraine Smith-MacDonald and her team recognised that morally injured first responders feel like mere 'meat in the seat' of an ambulance, fire truck, or police car (Smith-MacDonald et al., 2021).

The institution often fails to be a moral actor, becoming a moral hazard to those who work there. Institutions become places where the right thing may be punished while the wrong thing is protected through political and factional affiliations. Betrayal is no longer the exception, but the default position, as it is used to protect the institution from reputational harm. Institutional betrayals are rarely addressed or acknowledged internally; a media spotlight or judicial inquiry is needed to force change. Such externally prompted change results in a form of *mea culpa* (see Chapter 8). Institutions that resist all inquiry through the sheer weight of assets may gradually lose public credibility, thus becoming museums to a once-great past.

Museums

Philosopher and founding psychologist, William James, wrote: 'Most human institutions, by the purely technical and professional manner in which they come to be administered, end by becoming obstacles to the very purposes which their founders had in view' (James, 1909). Becoming a museum is not inevitable, but it does follow the trajectory of their administrative and cultural convictions. Real-life museums are the most authentic celebration of past glories. One is often presented with a dinosaur fossil, a knight's armour, or a cultural artefact forcefully obtained from a distant land. Museums scream, 'Look at all this good stuff from the past'. In the world of workplaces, whether churches, prestigious clubs, elite private schools, or hospitals, museums are now artefacts of history that are struggling to survive.

In a museum corporation, the core of the corporation persists, but its energy is depleted. Its primary activity is remembering what it used to be. Mission statements speak of bold change, but in practice, the organisation is risk-averse, ceremonial, and disengaged from or even at war with contemporary society. Museums still hold significant power and prestige, but have little purpose other than surviving their own extinction.

In this phase, betrayal comes through historical denial and opaque truth statements. The truth about past abuses is softened, revised, or erased and reframed as noble struggles. Those who speak honestly about what went wrong are seen as disrespectful troublemakers. Meanwhile, past successes are embellished and regularly rehearsed through patronage and ceremony. In a museum, moral injury is felt by those who remember the truth or have a lived experience of harm but see it denied. Ultimately, betrayal is not just about workers or their own mission; it betrays reality itself.

ITEMS FOR CONSIDERATION

- What kind of halo does your organisation have: assumed, manufactured, or genuine?
- What are the things you see in the shadow of that halo?
- What stage of evolution is your workplace in: movement, organisation, institution, or museum?

NOTE

1 This observation is made from Google Ngram graph of the use of the word halo. Ngram analyses the frequency of words from 155 billion words across 1.3 million titles. For more information on Ngram see, Michel, J.-B., Yuan Kui, S., Pinker, S., Nowak, M. A., Aiden, E. L., Aiden, A. P., Veres, A., Gray, M. K., Pickett, J. P., Hoiberg, D., Clancy, D., Norvig, P., & Orwant, J. (2011). Quantitative analysis of culture using millions of digitized books. *Science*, *331*(6014), 176–182. https://doi.org/10.1126/science.1199644

5 Organisational Betrayal

Leveraging the Resources of an Organisation to Crush Individuals

A true word.

> When a person speaks out from within an institution against that institution, they know reprisals are coming. Institutions carry it in their DNA, in-built self-protection, if you like. No one institution has a monopoly on reprisal; they all practice it to punish and deter.
>
> *Fox (2019), p. xi*

INTRODUCTION: GOING POSTAL

On the 6th of June 1965, at the age of 17, Hughie Noel Thomas commenced work as a Postman in the United Kingdom. Rising at 4 am, he delivered the mail daily through summer and Welsh winters on a gruelling 17-mile bicycle ride – but he loved it. By 1969 he and his wife were running what would become a series of different post offices over a period of 36 years. His faithful work for the Post Office was an act of service to the local communities in which he lived and worked. Postmasters and sub-postmasters were often not considered government officials, but were seen as trusted social hubs where connection, friendship, and service were central.

As the world increasingly digitised, the accounting system of the Post Office followed this trend. From 1999 onwards, accounts formerly done by hand were completed through a new centralised computer system known as *Horizon*. Sub-postmasters received minimal training for it, but were given a large, almost unconquerable manual and a phone line to assist them. Not long after the *Horizon* system was installed, Mr Thomas started having discrepancies reported in his accounts. He was told by the assistance line that he was the only one with these discrepancies. But, in fact, hundreds of sub-postmasters were having identical issues.

Thomas, like other sub-postmasters, had signed contracts with the Post Office in which they were responsible for accounting discrepancies and would be required to make up any shortfalls personally. In accordance with this agreement, Mr Thomas and many other sub-postmasters started paying a series of increasingly significant shortfalls that *Horizon* was reporting. The payments began as small amounts that Hugie and others withdrew from their personal savings; however, increasingly large debts continued to

DOI: 10.1201/9781003499725-7

accrue, so that many borrowed money or took second mortgages against these fictitious debts. All the while, hundreds of sub-postmasters were repeatedly told they were the only ones with issues. Debt and stress grew, but then the full weight of the Post Office was brought to bear against its sub-postmasters, particularly Mr Thomas.

After repeated concerns were raised, an investigation began, but it was only Mr Thomas and his fellow sub-postmasters who were investigated and punished. No systemic review of *Horizon* was done despite the growing number of issues. The first punishment for Mr Thomas came when the Post Office took the keys to his post office, sacked him, and removed the related accounting paperwork. It left Mr Thomas with no job and no evidence to fight the claims that he was not being diligent or honest in his accounting.

Then, in 2006, after 41 years of service, Mr Thomas was criminally convicted of false accounting over a falsely raised debt of £48,000. Despite over 100 letters in support from his community about his good character, Mr Thomas was imprisoned in Kirkham Prison for 9 months. During his imprisonment, Mr Thomas was declared bankrupt, and upon release, the Post Office attempted to stop his government pension. From 2005 till at least 2022, Mr Thomas could not find meaningful work or build a livelihood. Over 900 sub-postmasters were in a similar or worse situation, with at least one dying by suicide, whilst the Post Office divided and conquered its loyal and hard-working sub-postmasters.

Thankfully, one delightfully disagreeable sub-postmaster stood up against the might of an institution that had told him he was also the 'only one'. This brave soul, Mr Bates, has now been immortalised in a television series about his dogged fighting spirit that revealed widespread institutional betrayal by the Post Office and its senior leaders. A senior executive from Fujitsu in Europe (and creator of *Horizon*) testified that glitches and errors were known by Post Office managers 'right from the start'. At the same time, an independent report in 2003 showed that the system was 'clearly defective'.

What is certain is that the *Horizon* scandal was an undeniable corporate cover-up of epic proportions. Despite having awareness of the issues, the Post Office consistently denied public criticism of its operations. For example, in 2009, when *Computer Weekly* publicly broke the *Horizon* story, the Post Office replied, 'The Post Office *Horizon* system is extremely robust. It is regularly tested and scrutinised and we do not have any complaints from Sub postmasters'. At this time, key leaders would have been aware of the serious problems and numerous complaints, but their tactic was 'Deny, deny, deny!'

Approximately 700 prosecutions were initiated directly by the Post Office against sub-postmasters, with other charges being brought by the Crown Prosecution Service. As of early 2024, 103 of these convictions had been overturned, but at the time of writing, the legal ramifications for those leaders who pursued the sub-postmasters have not been fully realised. It appears that the legal consequences for Mr Thomas' fictitious crimes were much speedier than the justice that is yet to be meted out to the organisational leaders.

Mr Thomas has now had his conviction quashed, and the whole affair has been labelled by Rishi Sunak, former UK Prime Minister, as 'one of the greatest miscarriages of justice in our nation's history'. As for Hughie Noel Thomas, he reflected, 'You know, it was a betrayal, wasn't it?'. The Horizon scandal shows how leaders of institutions plan and enact the betrayal of those who work for them. During this saga, there was a notable lack of moral character among senior leaders, a protection of corporate halos, administrative violence, the use of gaslighting techniques, and numerous other underhanded tactics. The *Horizon* scandal is a good exemplar to spark our exploration of the mechanisms of organisational betrayal.

ENACTING BETRAYAL IN A WORKPLACE

The definitions of betrayal introduced in Chapter 3 are essential, but can leave us unsure about the actual methods used to betray a person. It is worth pondering the kinds of acts that might be utilised to breach deeply held expectations. To prevent betrayal, we must explicitly consider what nature and form pivotal acts of violation take in real life and in real time. It is crucial, therefore, that we explore examples of those actions and violations.

Betrayal is more than a destination; it is a perilous journey of many steps to the pit of hell. Betrayal in the workplace utilises a suite of grubby tactics that have been well researched across a range of contexts. In examining these actions, this chapter proceeds in three 'acts' plus an epilogue. Taking this chronological journey through betrayal will help one recognise the key terms used to defend organisational halos and enable betrayers to defend and attack.

Act One: A Catalyst Event

Domestic violence and abuse of the vulnerable are, to my mind, the most abominable forms of betrayal. To vow publicly in marriage, or privately through loving agreement, to accept the responsibility to cherish and protect a person and then act in direct contradiction to that pledge through intimidation, threats, harassment, assault, or murder is genuinely worthy of the centre of hell. Experts note that domestic violence often occurs cyclically. The cycle starts with a build-up of tension, a catalyst event that unleashes an explosion of abuse, followed by a period of relative calm before the cycle begins again.

The catalyst event should never be understood as an excuse for abuse that follows; however, that an event serves as a gateway to abuse helps us know how abuse and, in our case, betrayals habitually commence. Rather than a cycle, betrayal commences with a catalyst event that leads to a unitary downward trajectory towards accusation, investigation, and punitive action against the betrayed party. There are four primary trigger events for betrayal in the workplace: change, concern/complaint, corruption, and competition.

Change

Some hardy souls among us love change and can almost taste the positive outcomes that result from it. However, for most people, change is scary. The majority of people fear change because humans are more often wired to see what will be lost through the change. According to Kahneman and Tversky's prospect theory, people are loss-averse – we would prefer not to lose something we already have, even if we will gain something new and possibly better.

When I teach about grief and loss to emergency workers, I suggest that people grieve more than the death of loved ones, and more regularly grieve change in any context, including the workplace. The change and grief formula is simple:

Change = Loss
Loss = Grief

Any change, no matter how minor, in a workplace can be a catalyst for strong emotions associated with grief. Anger, fear, and uncertainty are strong emotions that help unleash acts of betrayal. For example, in Chapter 4, we described how the success of a start-up necessitates the change from an idealistic *movement* to an established *organisation*. There, we noted that the change can lead to clashes of priorities and strong emotions among staff. The grief associated with a monumental shift in a company's outlook impacts staff members' deeply held expectations. The change, even if necessary and obvious, can feel like a betrayal that evokes strong emotions often associated with grief.

The changes that led to the *Horizon* betrayal in the UK Post Office were not handled well. On the surface, the Post Office was aligning with necessary shifts in business and government to computerise and streamline accounting. While it was a significant change affecting approximately 18,000 post offices, it should not have been an occasion for betrayal. Had it gone smoothly, it should have been a time-saving change that benefited sub-postmasters. However, it was more than a system change – it was a costly and career make-or-break change for senior leadership. If seen to fail, that costly failure would be hung around someone's neck.

The original cost for *Horizon* in 1999 was UK £548 million. In 2025 that is approximately UK £1.03 billion, USD $1.32 billion, or AUD $1.96 billion. The UK Prime Minister at the time was pushing the implementation of this system to a watching nation. There was so much at stake that change *could not* fail. This pressure on senior leaders transferred the logic of *could not* fail to *had not* failed even when faults were repeatedly found. When money and ego are on the line in a must-not-fail situation, someone will take the blame when things go wrong.

If you are a C-suite leader, are you implementing a system that cannot fail due to cost or your personal commitment to the project? If so, what will you do when things inevitably go wrong? Will some innate self-protection kick in for you so that you blame those who complain or implement the system? It is of fundamental importance to consider the culture of the organisation you lead, as well as the culture you helped create and curate. So, first ask, 'does the organisation I lead have a system that seeks to find fault or improvements as a fundamental cultural value?'. If the answer is blame, the scene is set for organisational betrayal.

If you are a middle manager of a changing organisation, you are in a particularly perilous situation. You are not forcing change, but are enacting it with the frontline. You will receive the first notifications that all is not well, and then be required to communicate those issues to the higher leaders. Sadly, middle managers exist at just the right level to be blamed for any failure, or 'held accountable'. The more expensive the change and the higher the decision goes up the chain of leadership, the greater the risk for middle managers. When things start to go wrong, your career and, more importantly, your integrity will be on the line. Will you seek resolution of complaints or bury them? Let's turn to the notion of complaint as a catalyst for betrayal.

Concern/Complaint

Dr Nathalie Martinek has become an avid researcher in narcissism and scapegoating in the workplace (see Chapter 7). She describes a flow of action when one raises concerns in the workplace:

> In status-driven workplaces, it doesn't take much to trigger the Scapegoating Playbook.
>
> - You ask a question about a process everyone pretends to understand.
> - You request details that were conveniently left vague.
> - You offer a perspective that sits outside the group's carefully curated consensus. You're not trying to disrupt anything. You're doing what competent people do. But the reaction tells another story:
> - Meetings suddenly feel heavier
> - Praise dries up
> - Decisions stall for "timing" or "political" reasons
> - Your approach becomes a topic whispered, not discussed
>
> Why? Because your actions quietly shifted the balance. You revealed a gap, an inconsistency, a flaw in the system someone was invested in protecting.

Sadly, Dr Martinek understands that in many industries, those who raise concerns are viewed as troublemakers or incompetent operators. Punishment is used to deal with these detractors, but it is often employed in a passive-aggressive manner – praise dries up, and your name is whispered. Martinek argues that this process is a broken system that operates exactly as intended – to suppress dissent. When this process occurs, everyone else sees it and ceases to raise concerns or complaints, fearing they too will be targeted.

In industries where errors and complaints are suppressed, there is a lack of safety on the preventive side of incidents, which affects day-to-day operations. On the preventative side, Harvard leadership Professor Amy Edmondson labels the lack of safety as a lack of psychological safety, which she defines as a 'shared belief held by members of a team that the team is safe for interpersonal risk taking' (Edmondson, 1999). Raising a concern is an interpersonal risk that must be mitigated to create psychological safety. When people are afraid to speak up because the risk of personal consequences is high, possible safety incidents are neither known nor prevented, decreasing physical safety.

The aviation industry has numerous rules governing aircraft maintenance, in-flight safety, communication, innovation, crew rest, and many other aspects. Many aviation veterans will tell you that these rules are written in blood. That is, an adverse incident occurred where someone was killed or injured, and the ensuing investigation recommended new guidelines to improve safety. Obviously, there are better ways to create aviation safety guidelines than aircraft crashes. That is, near misses or the expression of concern about a process or equipment before a collision occurs are vital learning experiences, allowing aviation experts to learn as much as they would from a crash, but without the loss of life. Truly a win-win!

However, to encourage people to report near misses or safety concerns, there must be a culture that views complaints and errors as foremost opportunities to learn and improve, rather than as opportunities for blame and punishment. If errors and complaints are punished or suppressed in the aviation industry, more people must die to learn what could have been learnt through non-fatal means. It is challenging for many leaders in non-critical workplaces to understand the concept that complaints and mistakes can be lifesavers.

It is important to pause here and ask some critical questions about your workplace. If you are a leader, how do you react to error or complaint? Do you use retribution and punishment as an example for others? Do you perhaps feel threatened by their behaviour? I now have a key diagnostic question to ask prospective employees: 'How do you respond to criticism and complaints by co-workers and subordinates?' How do you answer that question? Lives might depend on your answer.

Perhaps you are the other kind of leader who acts with curiosity and compassion. These leaders want to deeply understand core issues, and act in a way that is inquisitorial rather than adversarial – they want power that comes from knowledge, not power that suppresses understanding. For the leader who wants to respond well to complaints and concerns, it can be hard to articulate their approach, let alone adopt it in a systemic way. Chapter 10 discusses a way of thinking about improvement and learning in critical industries through Just Cultures, which rely on psychological safety and systematise them into an organisation-wide relational network of trust, learning, and forward-looking accountability.

Corruption/Criminality

A catalyst for betrayal is often the revelation of corruption, criminal activity, or inappropriate behaviour by people of repute. This catalyst is a variation of raising a concern or complaint about a programme in but it is intrinsically personal. Therefore, the response is personal and often more visceral.

We previously mentioned the most recent findings of the Edelman Trust Barometer about the rise of grievance. In 2025, Edelman also noted people are increasingly thinking with a 'zero-sum' mentality. In this mindset, there is no such thing as a win-win; one only wins if one's enemy loses. The reverse is also true, where an enemy's gain is considered a loss for you. This zero-sum mindset fuels political polemics and escalating geopolitical tensions. It prevents us from recognising that individuals with opposing viewpoints can also achieve positive outcomes. Conversely, we are blinded to the fact that people or groups that have halos are capable of poor behaviour.

Corrupt individuals are often gregarious, successful, generous, and highly likable. They are 'good' people who do bad things. The truth is that corrupt individuals know how to manipulate systems and people so that a loyal group of emissaries defends their interests. These people with a veneer of goodness seem to get away with any manner of crime, partly because in-group friends vehemently jump to their defence.

Revealing their corrupt or criminal behaviour rarely elicits an immediate confession of illicit behaviour. Instead, it often leads to betrayal and counteraccusation in business, government, the media, and even sport.

Lance Armstrong was a former champion cyclist. He won an unprecedented seven Tour de France titles, defeated cancer, and founded the enormous cancer charity *Livestrong*. In 2003, Armstrong was on top of the world, with a loyal cadre of friends and celebrities to defend him. That year, his former soigneur, Emma O'Reilly, provided information to journalists about Armstrong's illegal use of performance-enhancing drugs.

Armstrong labelled O'Reilly an alcoholic and a prostitute to denigrate her character. One of the journalists she spoke to was sued by officials of the international governing body of cycling and labelled a 'little troll' by Armstrong. Two things are noteworthy. Firstly, most people, including myself, stayed on the Armstrong bandwagon for longer than they should have. And secondly, it was a zero-sum game for Armstrong. Armstrong didn't just have to win; his accusers had to hurt for their revelations.

Those who accuse star performers of poor behaviour are often portrayed as disloyal and then cast as the betrayer. Those who act as whistle-blowers rarely 'win'. Consider that O'Reilly was accused of being an alcoholic and a prostitute in 2003 and was vindicated only in 2013 when Armstrong publicly admitted his sins and apologised. History does seem to collapse time, but for O'Reilly, it was a decade of pain and distress. Many who reveal corruption wait even longer for vindication, if they receive it at all.

A sign that betrayal is in play is when those vehement, even offensively stated, denials of wrongdoing are made. When the accuser is strongly and instinctively attacked, Shakespeare's aphorism 'The lady doth protest too much, methinks' is worth considering. The more vehement the denial, the more likely the accusation is to have a large amount of truth. Alternatively, when accusations are replied to with corporate jargon about 'miscommunication' or 'misunderstanding', treachery may well be afoot. When the response to an allegation is 'we are sorry that this person has been offended', you can almost smell the fumes of passive-aggressive denial.

It is worth asking who the pin-up, stand-up, all-round high achievers are in your organisation. Is it conceivable to you that they could do the wrong thing, or has so much been invested in their persona? In revealing the organisational coding error that enables organisational self-protection, it is even more critical to ask: if a revelation of wrongdoing by the star was proven, would they be immediately cancelled, stood down and fired for admitting their fault? If the answer to this question is a resounding 'yes', you have located their motive to attack those who reveal corruption. If restoration were an organisational code, then part of the motive for attacking detractors would evaporate. The zero-sum mindset that casts people as heroes to be lauded or zeros who need dismissal must itself be dismissed.

Competition

The final catalyst event for betrayal is the introduction of competition. In most workplaces, managers and employees compete for money and glory, with threats to our income stream and workplace status being met strongly. In most organisations, there is a limited amount of money and glory, so when a rival department or a new, competent team member is employed, they are often perceived as a threat rather than an asset. Competition can bring out the best in us, motivating us to strive for excellence. However, with a growing zero-sum mindset in society, competition increasingly brings out the worst in people.

The competition that leads to betrayal can be personal or team-based. On an individual level, it occurs when a new and competent employee is introduced. This interloper may be perceived as a threat to other team members or the manager who oversees them. When they ask probing questions to improve procedures or outcomes, they are first met with resistance and then increasingly undermined by those who feel threatened.

It is worth noting that the talented new hire may not, at least subjectively, be competing with others. Dr Martinek notes that because this person is competent, they feel they are simply doing the best they can. Their talent enables them to do it better than others, and therefore poses a threat. Because they intend to do their best rather than compete, it may come as a surprise that they are viewed as a threat rather than an asset – a typical reaction among those who have been betrayed.

Competition can also come from other teams competing for funding and career paths for the team leaders. Recently, I witnessed competitive instincts come out in two rival government agencies. The agencies had been part of one large agency that was divided into two after an inquiry found that the larger agency had not met expectations. The competition the divide created was not pretty. A senior manager of one agency described the reallocation of funding as a 'knife fight', while other managers described themselves as 'children of a divorce'.

There are many admirable and hard-working people in both these agencies. However, these statements reveal that the political dissolution of the original agency has left many workers deeply hurt and even betrayed, while others are up for a fight. These fight-or-flight instincts are evident in competition that manifests as the lack of sharing of scarce resources and the distinct feeling that if one agency loses, the other wins.

Your organisation will have a marketplace rival with whom you compete for market share, but they are probably not your most significant threat. A bigger and often unconsidered threat to your operational performance, reputation, and staff well-being is internal competition. Therefore, if you are in the C-suite of your organisation, it is worth considering if you have explicitly or tacitly set departments up against each other. It may seem that this competition will yield excellent results, and in the short term, it might. However, a new generation of younger workers is not happy and is willing to quietly quit or make workplace psychological injury claims against this unhelpful dynamic. And good on them, I say! Chapter 10 explores how this competition might be changed to complementarity.

Act Two: Circling the Wagons

For every action there is a reaction; however, with acts of betrayal, it will not be an equal and opposite reaction. The chain reaction of betrayal stemming from a catalyst event will be disproportionate. Change, concern, corruption, and competition are all trigger events that cascade into a reaction, forming the next stage of workplace betrayal. The first action is to defend the turf held by those in power in the movement, organisation, institution, or museum, which is believed to be under threat. The first line of reaction is an instinctive circling of the wagons to deny the credibility, substance, or seriousness of the difficulty being addressed.

In John Grisham's 1995 novel, *The Rainmaker*, an insurance claimant, Mrs Black, had her reasonable claim repeatedly denied by the Great Benefit health insurance company. The matter eventually finds its way to court, where it becomes obvious that her treatment has been deeply unfair. A courtroom crescendo rises when a final letter of claim denial is embarrassingly read in open court with the final words, 'You must be stupid, stupid, stupid, stupid!'.

The first tactic of defence is simple and repetitive: deny, deny, deny. While we must remember that Great Benefit and Mrs Black are fictitious, Chapter 9 will detail the grievance against real-world insurance companies, which many believe mirror Grisham's artistic courtroom portrayal in real life. Furthermore, two Royal Commissions into large Australian institutions have demonstrated how organisations often defend and deny as a tactic to maintain their reputations.

The Royal Commission into Defence and Veteran Suicide was held in 2021 and 2022. It heard evidence from 340 witnesses, conducted 900 private sessions, and made 122 recommendations. As the final report was about to be published, Assistant Commissioner Dr Peggy Brown AO slammed the Australian Defence Force (ADF) and the Department of Defence and Veterans' Affairs. Brown said the ADF was incapable of holding the public's trust to solve issues of suicide by their members. The reason she gave was that, 'We have seen too often that Defence's default position is to **defend Defence**, and too much evidence of "busy work" at both Defence and DVA giving the appearance of implementing change, when in reality they have only served to perpetuate old ways of operating' (my emphasis) (Royal Commission into Defence and Veteran Suicide, 2024). The ADF and other government agencies 'stymied and stonewalled' the commission, hid unflattering evidence, and made superficial changes to make it look like positive change was occurring.

The Royal Commission into institutional responses to child abuse heard evidence from around 1,200 witnesses and had 8,000 private sessions that produced 189 recommendations. While many government and community organisations were revealed to have engaged in cover-ups of abuse, it was the reaction of churches that was particularly galling. Various church bodies were, much like the ADF, appropriately eviscerated for their treatment of those who spoke out.

Religious institutions, particularly Christian ones, had been ubiquitous and mostly above suspicion of wrongdoing. Around religious leaders and organisations was a halo of unassailable light. Churches jealously guarded their reputation as doers of good, partly by doing good things, partly by appearing to do good, and lastly by quashing any accusations of wrongdoing made against one of its agents. As these institutions evolved into cultural museums, they instinctively acted to defend themselves against allegations.

Peter Fox, a Detective Chief Inspector of police, acted as a whistleblower against church abuse and was the subject of controversy as a result of his stand. Fox wrote of his treatment by some of his colleagues in the police and the battle against the church in his region. Both institutions had a desire to defend their interests rather than pursue justice. This abuse of power in the cover-up of wrongdoing was not limited to the Australian church. In 2009, a similarly extensive investigation in Ireland reported the churches'

> preoccupations in dealing with cases of child sexual abuse, at least until the mid-1990s, were the maintenance of secrecy, the avoidance of scandal, the protection of the reputation of the Church, and the preservation of its assets. All other considerations, including the welfare of children and justice for victims, were subordinated to these priorities.

Again and again, this tactic of denial, delay, diminishment, and dismissal is used by movements, organisations, institutions, and museums. Denial is a subtle but effective starting point, done to destabilise and discredit those who bring issues to light. For those who raise concerns, there can be a feeling of disconnection and disbelief in the denials of something they have witnessed. The organisational denials are crafted to cause doubt about the accusation and the accuser.

One tactic used in the denial phase is gaslighting. The term originated in the 1938 British play 'Gas Light' by Patrick Hamilton, which later became the 1944 movie 'Gaslight'. The plot sees a man manipulating his wife by dimming the gaslights in their home, but he denies that any difference had occurred. As a result, she was made to doubt her ability to process information and the conclusions she draws.

Gaslight informed the 1967 movie 'A Guide for the Married Man', which depicted how men could cover their adulteries by making their wives believe they were confused or not able to understand the real situation. The tactic was to persistently deny, even in the face of the most obvious evidence of wrongdoing. The power of denial grows each time it is done.

The phenomenon of denial is the first part of a powerful paradigm that explains the process of organisational responses to concerns and allegations of corruption. Deny, Attack, Reverse Victim, Offender (DARVO) describes a methodology employed during the process of betrayal.

DARVO

The term DARVO was first coined by psychologist and researcher, Dr Jennifer Freyd in 1997. In a paper on betrayal trauma, DARVO was introduced as a speculative framework that may help describe organisational betrayal. Freyd hypothesised that denials could be assessed by the style of their delivery: 'If an accusation is true, and the accused person is abusive, the denial is more indignant, self-righteous and manipulative, as compared with denial in other cases' (Freyd, 1997). This is something I can imagine Emma O'Reilly understood well when she accused Lance Armstrong of cheating.

Kurt Gray and Daniel Wegner have researched two regular methods used to defend and deny responsibility when questions are raised about immoral or unsafe practices (Gray & Wegner, 2011). The first tactic is to acknowledge the issue, but then lean on the halo of the organisation's history or heroic efforts in the face of any allegations. However, they argue this tactic is rarely effective.

A better denial tactic is to play the victim. When a violator suggests they are somehow a victim, even of something unrelated, they are blamed less for their actions by the public. Who hasn't heard of a celebrity or CEO bringing up past sufferings, or the perennial claim of suffering from a mental health issue to avoid culpability? The denial of wrongdoing and the reversal of victimhood is often interspersed by attack, which is the third act of betrayal.

Act Three: Turning the Tables

Lance Armstrong labelled Emma O'Reilly a prostitute and alcoholic, and attacked any other accusers by leveraging his substantial legal and business might to crush them financially. The Post Office brought charges against innocent workers and even sought to have their pensions cut off. The ADF engaged in 'administrative violence' through which due process was weaponised to ensure whistleblowers would be put in their place. The beauty of this administrative violence is that the abusive organisation can point to all they are doing to address the concern, meanwhile using those processes to outstay the accuser.

Outright attacks, such as those of the Post Office, are effective; however, many attacks possess the ironic combination of subtlety and power. Often, the attacks are a form of psychological operations (PsyOps) in which passive-aggressive tactics or linguistic contortions, such as gaslighting, are used, rather than full-frontal attacks favoured by people like Armstrong. What Armstrong and the Post Office had in common is that their substantial power came from having substantial resources. These resources can be used to prolong any investigation, legal action, or due process so that 'David' is unable or unwilling to keep pursuing the matter against a corporate or celebrity 'Goliath'.

PsyOps

PsyOps have a military origin as tactics to manipulate the perceptions of target enemies or communities. However, at least one military organisation, the ADF, managed to shift those tactics towards its own staff. Administrative violence was recognised and detailed by researchers on mental health and wellbeing subject matter experts (Wadham & Connor, 2022). Their submission to the Royal Commission into Defence and Veteran suicide detailed this tactic and was supported by numerous testimonies. Administrative violence includes 'actions that are undertaken within institutional rules that are designed to further traumatise, belittle and/or restrict access to support and/or redress'.

Administrative violence uses bureaucratic and disciplinary systems to punish, isolate, or silence those who report wrongdoing or show signs of psychological distress. Rather than receiving support, personnel who raised concerns were often met with petty or demeaning tasks designed to humiliate, such as being ordered to paint rocks. Many witnesses reported being denied leave, refused training opportunities, or missing out on favourable postings, and their career progression was deliberately and deniably stalled.

Like the Post Office, the ADF was not satisfied with simple punitive actions and ensured individuals were subjected to retaliatory disciplinary actions or medically discharged while receiving treatment. Discharge not only removed an accuser's livelihood, but it also effectively cut them off from the care and community they now needed more than ever.

The impact of this administrative violence was profound and quite often more damaging than the original event. For many, this violence constituted a second and more profound betrayal. It stripped away their identity, purpose, and support networks, leaving them psychologically exposed and socially discarded. The fear of experiencing similar treatment deterred others from speaking up, perpetuating a culture of silence and stoicism. In one especially harrowing case, a former Navy officer who reported abuse was subsequently bullied, medically discharged, and subjected to aggressive legal cross-examination during the Royal Commission, events that nearly drove them to suicide. Such stories reveal how administrative systems can become tools of betrayal that amplify harm and actively undermine trust, safety, and well-being in the workplace.

Understanding and detailing administrative violence as a process undertaken within an organisation denies the reality of what is happening. Let's be clear, organisations and their culture are merely functions of the people who work in them (see Chapter 4). The sub-postmasters and ADF staff were betrayed by their managers, commanders, and complicit underlings. The power wielded by commanding officers and executives often enabled reprisals to proceed with little hope of justice.

Unlike the individual victims of administrative violence who suffered immediately, no individual commander or leader in the ADF has yet faced legal sanction for their actions. At the same time, convictions against post office managers may have to wait until 2028, 22 years after Hughie Thomas was falsely convicted. The injustice of administrative violence is fast-acting against those who raise issues, while the justice for perpetrators is slow-moving at best, and non-existent at worst.

A subtle twist that DARVO introduces is the move to make the whistleblower into the real villain – used with the subtlety of a sledgehammer by Lance Armstrong, who labelled O'Reilly as a drunk and a whore, but more often it is subtle. Particularly in *movements*, there can be emotional blackmail Cindy received, such as, 'you wouldn't want to ruin our good work by pursuing this revelation?'

Through the utilitarian logic, the act of raising an allegation against a haloed organisation or its leaders is enough to allow the accuser to be portrayed as the one causing harm. The leaders or HR workers who are seeking to defend and deny will feel justified in playing off a seemingly small indiscretion against the greater missional goal. For the one raising the allegation, it may be enough to make them stop their complaint, but it will almost certainly cause a moral injury.

Don't fall for this kind of emotional blackmail. It is subtle but powerful. If you hear this logic being used against you, watch yourself carefully. Save every email you send and receive about the matter, and start looking for a new job. The organisation is rotten, and you will be targeted if you continue. Worthy organisational mission outcomes are not worth compromising your integrity for, or sacrificing your health when organisational self-preservation takes precedence.

More could be said, particularly about how whistleblowers are often portrayed as villains, but space is limited. Chapter 10 will examine what good looks like through an examination of institutional courage recommended by Freyd and her team. Before this chapter concludes, it will be beneficial to note the power disparity that exists in many organisations. Raising concerns against large *organisations* and *institutions* is dangerous because size matters.

Size Matters

The classic biblical account of David and Goliath has become a cultural meme for the smaller but righteous warrior of truth winning their battle. However, the story is the exception rather than the rule. Those Royal Commissions and the *Horizon* investigations revealed hundreds of innocent victims, as Goliath had vanquished so many. Therefore, as a whistleblower, you should not expect the righteousness of your cause to be a guarantee of success, as organisations and leaders are willing to throw their weight around.

What a whistleblower often experiences is death by a thousand policies. Once an organisation has become so large that it can be considered an institution, it develops a form of hubris that it is untouchable and self-authenticating of its moral integrity. Hubris is often the key ingredient for organisational betrayal because leaders believe their station and their cause are beyond questioning, let alone by the staff who work there.

Many large organisations have so many policies that they can never be fully obeyed. In the process of maligning a whistleblower, leaders will task 'investigators' to find the policies you have breached. Many organisations have both the money and will to reallocate resources to crush dissent. You might complain about serious breaches of safety, or psychological abuse by a senior leader. Still, the resources of the corporation will find that your fire safety training wasn't completed on time, or your work phone was used for a personal call. The abuser becomes the victim and you are painted as the lazy employee out for vengeance.

Model litigant literature recognises that power imbalances exist and demands that Goliaths do not win through their ability to outpay and outgun those who raise complaints. Corporations acting as model litigants seek not to win through attrition; instead, they seek to uphold justice, fairness, and public integrity when dealing with individuals or small entities. The Australian Government has outlined recommendations for how agencies are to act as model litigants (Office of Parliamentary Counsel,

2017). In light of their principles on model litigant behaviour, a reliable leader will ask if their organisation,

- Acts truthfully (see Chapter 6) and fairly in handling claims and litigation.
- Avoids causing unnecessary delays that you can absorb but individuals cannot.
- Does not take advantage of the knowledge that a whistleblower lacks resources.
- Does not rely on technical defences unless absolutely necessary.
- Keeps your own costs to a minimum and a reasonable level.
- Considers alternative dispute resolution (e.g. mediation).
- Apologises where appropriate.
- Deals with claims promptly and not contest liability where clearly established.

If your organisation fails to comply with these recommendations, then it is neither fair nor moral in its treatment of workers or clients. One day, such hubris will come to light, and a moment of *mea culpa* will result in the change you have resisted.

EPILOGUE: THEIR WORDS BETRAY THEM

Frequently, early evidence of organisational hubris and impending cover up and betrayal is evident from the words used by leaders. Below is a list of phrases that act as markers of organisational hubris. Leadership professor and researcher Dr Vita Akstinaite and her team analysed indicators of hubris in the language used by CEOs (Akstinaite et al., 2020). They noted the use of power words, negative emotions, and, importantly, specific pronouns as indicators of hubris. Pronouns, irrespective of their gendered elements, can provide strong clues that your manager is developing the hubris that underlies betrayal.

When leaders speak of company achievements with the use of 'we' (e.g. 'we've hired lots of local people'), there is an evident blurring of organisational and leadership roles. Those company achievements will be defended with hubris and deep personal animosity. Alternatively, hubris-filled leaders may use second-person singular pronouns (you/yourself/your) when describing how they have risen above others. For example, when they speak of overcoming hurdles, they may say, 'when things aren't going right, you have to ask yourself…'. Finally, hubris-filled leaders may repeatedly use the impersonal pronoun 'it' rather than naming a project they have worked towards.

In place of hubris, it is suggested that curiosity is a more effective and helpful leadership stance. This chapter will conclude by detailing the signs of betrayal in a game of hubris bingo. In the left-hand column of Table 5.1, you will note phrases that arise when change, complaint, corruption, and competition are brought up. They depict a form of leadership and organisational hubris, warning signs that may lead to betrayal. If you hear them too regularly, start preparing your CV before it all comes back onto you.

In the right column are phrases of curiosity, which, if actioned, are positive indicators of psychological safety and of Just Cultures in play. Unsurprisingly, they

TABLE 5.1
Hubris vs Curiosity

Statements of hubris	Statements of curiosity
You must play the game.	The teamwork we expect has not been present.
You don't understand the deeper issues.	Can you help me to understand the deeper issues?
Addressing this issue might make us look bad and derail the good work we do.	This is an issue of integrity. How can you see start addressing this underlying issue?
Your expectations are too high.	There appears to have been a breach of the trust you placed in us.
You need to stay in your lane.	I recognise that there are cross-department issues that should be addressed.
The people involved have, a) good intentions, b) given so much in the past.	No one here is above questioning. How might we start to resolve these issues?
Sorry… that you are upset.	Sorry that I/we have hurt you.
Sorry, but…	Sorry. No ifs and no buts.
We couldn't possibly change like that.	What do you think we could do to address this?
We don't have the money to address this.	How do you think we could reallocate our assets to solve this issue?
We are better than… a) our competition b) we once were.	This is important and reminds us we need to keep improving each day.
These issues are far too complex to be addressed easily.	I understand the issues to be, am I missing anything?
Leave it to me.	How would you like to be involved in resolving this?
You are the only one.	Your concern is important even if it is the only one.
We are the experts in this.	Your contribution is important/helpful.
It's out of my hands, really.	When would be good to follow up on this?

regularly finish with a question mark. The right kind of questions are curative of toxic organisational cultures, while the statements of hubris are often acts that are described by the ancient art of paltering.

ITEMS FOR CONSIDERATION

- What types of catalyst events have you seen bring unsavoury organisational practice out in your workplace?
- When you are faced with a complaint, concern, corruption allegation, or competition what is your default setting? How do you respond to challenges such as these?
- What phrases of hubris have you used yourself? How might you turn hubris into curiosity?

6 Paltering
Halving the Truth

A true word.

> *Administrative deceits, in politics or elsewhere, falsify the conclusions drawn from a fact!*
>
> Col. Ardant Du Picq

> *The truth shall set you free!*
>
> Jesus Christ

> *What is truth?*
>
> Pontius Pilate

INTRODUCTION: POWER AND SHAME

In 1995, a 22-year-old college graduate landed an unpaid internship at possibly the most prestigious organisation in the world. Within 6 months, that unpaid internship became a paid internship, likely the start of a young woman's dream career. Despite this promising start, from 1995 until 1997, the most senior leader in the organisation, a 49-year-old married man with immense power and prestige, engaged in inappropriate behaviour with the 22-year-old. This behaviour included the performance of oral sex in his office. It was a curious and dangerous leveraging of power differentials. When this relationship was revealed, there was too much at stake for the leader to address the issues with clarity, or what we might call the whole truth.

As a way of minimising the seriousness of the activities, the male leader called what happened 'physical acts' rather than 'sexual acts'. This distinction was one small linguistic step for this man, but a giant leap from the truth for the office that he held. That man was President Bill Clinton, and the intern was Monica Lewinsky. Those old enough to remember the 1990s will remember the Clinton and Lewinsky 'interngate' scandal and the public affairs nightmare it was for Clinton, but may not appreciate the way words were manipulated nor the long-term impact of the saga.

DOI: 10.1201/9781003499725-8

When the story broke in the worldwide media, President Clinton said, 'I did not have sexual relations with that woman, Miss Lewinsky'. Clinton doubled down on that claim, saying, 'There is not a sexual relationship, an improper sexual relationship or any other kind of improper relationship'. At some point later, Miss Lewinsky produced a semen-stained blue dress she had worn during one encounter, which became a factual relic that made Clinton's hard-line denials less and less believable. Still, Clinton engaged in all manner of verbal gymnastics to avoid the consequences that come when people in power are found out.

When Clinton stated, 'There *is* not a sexual relationship' between him and Lewinsky, it was not strictly a lie. After the blue dress emerged and during impeachment proceedings against him, Clinton argued that he had not lied in that statement about a sexual relationship being non-existent. He argued that because all activities, sexual and/or physical, had ceased at the exact time he made the statement, his use of the present tense 'is' rather than past tense 'was' contained the truth. We might wonder if Clinton told the truth or had just offered some spin to avoid consequences.

The career trajectories for the two parties after the scandal broke could not have been more different. Clinton successfully fended off an impeachment and left office with the highest presidential approval ratings in US history. He largely maintained his status, prestige, and wealth and has moved on to a successful career in publishing, public speaking, and philanthropy. Lewinsky, on the other hand, struggled to find work or lead a normal life in the following 20 years. She has never entirely escaped her very public and perennial shaming.

The might of the Office of the President of the United States and the subtleties of the English language won the day. Truth itself came out badly bruised – a battering from which it has barely been able to recover. Truth is a difficult concept to master, and even harder to tell consistently. Telling the truth may cost you your job, your reputation, and substantial sums of money, but not telling the truth may cost you your soul.

To minimise the temporal cost of truth, institutional leaders often fall short of telling the truth, the whole truth and nothing but the truth. In the act of workplace betrayal, lack of truth is frequently a core part of the process. Falling short of telling the truth does not mean leaders *lie* to cover over their sins or those attributed to their organisations. There is a significant gap between lying and telling the truth, which the ancient word 'paltering' aptly describes. Paltering involves manipulating the gap between truth and a lie to one's advantage. The main point of this chapter is that truth and lies are not categorical but are spectral realities. The spectral nature of truth provides wiggle room for linguistic trickery that can protect organisational interests while dismissing complaints and silencing those who bring concerns to light.

Clinton may not have lied per se, as fellatio is a physical act as much as a sexual one, but it would be hard to argue that he told the truth. When he used the present tense 'is', he did not lie, but it had the effect of misleading the world through a convenient palter. To better understand paltering, let's first examine the nature of truth.

WHAT IS TRUTH?

Truth is a commodity that humans seem to value highly because they have examined its nature and extolled its virtues throughout history. Philosophers such as Aristotle, St. Augustine, and Immanuel Kant have written extensive works on the subject. In courtrooms around the world, those giving evidence are asked to make a promise to tell 'the truth, the whole truth, and nothing but the truth'. Good parents teach their children to 'tell the truth', and when they grow up, these children look for partners who, among other attributes, will tell them the truth. Truth really matters, but what is it?

TRUTH, LIES, AND THE GREY BETWEEN THEM

Sometimes, it is helpful to examine the opposite of something as a way to understand the thing itself better. Therefore, we will begin by examining what many consider the opposite of truth – lies. When a legal scholar and a political economist come together to discuss lies, it is worth listening. Professors Frederick Schauer and Richard Zeckhauser collaborated on numerous papers, but their paper on paltering is a fascinating insight into truth, lies, and what inhabits the space in-between. They note that lies involve:

1. intentional communication,
2. that a person knows to be false,
3. that the thing communicated is in reality false and,
4. produces the effect of having the receiver believe or be likely to believe in the truth of something that is not in fact true (Schauer & Zeckhauser, 2007).

While falsity is central to a lie, the relationship is just as central. A lie is not something kept to oneself – that is a delusion. Lies are something one tells another person, meaning a lie is as much a verb as it is a noun. Lies are social actions of betrayal as much as they are epistemological categories.

Another critical part of lying is that the realities behind the utterance are falsifiable. It is possible to detect a lie from directly contradictory evidence and then sanction someone for the deception. For example, the offence of fraud involves a lie which can be proved by contrary evidence and then punished at law. Courts also punish those who bring their lies into court through the crime of perjury. Organisations have codes of conduct that strictly forbid lying. The telling of lies has practical difficulties – people can be called out and punished – so direct lies are a bad tactic in defending organisational halos.

Schauer and Zeckhauser argue that the act of truth-telling is just as important, if not more so, than the truth itself. Truth kept to oneself is secrecy, and sometimes it is a dangerous secrecy. For example, to withhold the truth of an impending disaster is more than negligent; it is culpable. Schauer and Zeckhauser regard truth-telling as:

1. the sincere assertion of a proposition that is both
2. believed to be true and
3. actually true, and

4. where the assertion either creates or reinforces a belief by a listener in a true proposition.

Truth and lies, it seems, are simply the opposite sides of the same coin. However, Schauer and Zeckhauser were even more intrigued by the ability to neither lie nor tell the truth with the same utterance. They write, 'numerous statements are, intriguingly, neither lies nor truth-tellings. Rather, they make up a universe of morally and socially problematic statements and actions in which one or more of the elements of the true lie is missing, but one or more of the elements of authentic full-bore truth-telling is missing as well' (Schauer & Zeckhauser, 2007). That is, truth and lies are not on opposite sides but on either end of a spectrum, with the space in between filled with opportunities for paltering.

Paltering, to the duo, falls short of a complete lie in two ways. First, a palter may not be factually incorrect. For example, fellatio is a physical act, so not saying it is a sexual act is a creative use of language; it is a palter. Secondly, a palter seems less harmful than a lie. Unlike outright lies, one often escapes sanction for paltering. Indeed, Clinton prevailed in his impeachment trial, but may not have if he'd said, 'I have never met that woman', a statement that is falsifiable by video evidence. Paltering, then, is much preferred if you do not want to tell the whole truth, and don't want to get caught out in a lie.

Truth and Facts

Before examining the use of paltering in betrayal, it is helpful to explore a word often incorrectly conflated with truth – *fact*. Facts are objective events or objects that are verifiable by our senses. For example, that the Space Shuttle Challenger exploded on 28 January 1986 killing all seven astronauts on board is a fact. Facts are vital but only get us so far. As meaning-making and sense-making creatures, we also want to know the *truth* about events like the Challenger disaster, and ask questions like: 'What caused it?', 'Who is to blame?', 'Could another Space Shuttle explode?' (it did).

Generally, facts are either physical (existing) or historical in nature (having either existed or happened). Truth, on the other hand, is an epistemological category that makes sense of facts and has relational consequences involving trust. While truth can be difficult to establish outside of facts, facts do not necessarily lead to the telling of the truth.

French mathematician Henri Poincaré is reported to have sardonically quipped, 'Science is built up of facts, as a house is built of bricks; but an accumulation of facts is no more a science than a heap of bricks is a house' (as reported in Altman, 2012). That is, facts are relatively easy to locate, but a degree of sophistication and dedication is required to corral facts in a manner so that truth may prevail.

Much like bricks create a wide variety of houses (or piles of rubble), when different people handle facts, different truths appear to emerge. For instance, prosecution and defence lawyers interpret and present facts in different ways to persuade juries to find a verdict of guilty or not guilty. Multiple apparent truths frequently arise because people distort, manipulate, discount, misunderstand, hold biases and preferences, or lack full access to the relevant facts.

Outside the courtroom, different versions of the truth present an opportunity for corporate press releases and HR doublespeak to spin the narrative, while avoiding the consequences of telling the truth or lying. But as Poincaré infers, truth is more complex than just access to facts – it is about how those facts are constructed to build truth.

Schools of Truth

There are several broad schools of thought that describe the relationship between truth and inputted facts, as well as the active output of an ethical and truthful life. Four dominant ones include correspondence, coherence, pragmatist, or constructivist approaches.

The correspondence theory states that truth must correspond with the evidence as it is discovered. Truth, therefore, is only derived from empirical evidence; however, at times, correspondence proponents really do just make a pile of bricks, not a house.

Coherence theories prioritise the internal consistency of the truths we hold over whether they are provable from the facts, at least initially. The reasoning process is that of abduction, which seeks not to prove anything, but create the best explanatory fit among a web of beliefs we already hold. Doctors commonly undertake the process of abduction as a first step towards finding a reliable diagnosis. They first form a professional opinion that aligns the range of symptoms with their current knowledge to guide further investigation, such as X-rays or blood tests.

The pragmatist approach favours whether a truth 'works' to achieve an aim. For example, if my hammer puts screws into timber well enough for my liking, my truth is that the hammer is a screwdriver – it works. In the workplace, pragmatists often shortcut standard operating processes that are ineffective and inefficient for a method that actually works.

Finally, the constructivist approach is almost the reverse of the correspondence theory. It holds that humans construct truths contextually as individuals in a particular society, with equally valid ways to discover truth that then shape our view of facts. For example, the manager and the factory worker will construct different truths of the current union dispute regarding wages and working conditions. Or fans of opposing football clubs will come to varying conclusions around a referee's decision by constructing narratives with those who also support their team. For more radical constructivists who lean towards relativism, truth is primarily a construct of your context, with evidence somewhere between a secondary concern or an obstacle to finding truth.

These four schools might be well explained by different readers of this book. Those in the research world will continually be saying, 'Where is the data to support that statement?' Or for the hardcore researcher, 'Was there a randomised control trial?' These people are correspondence people, and to satisfy them, I have included references to research so that they can see there is supporting evidence for this work, but I have not included hard data everywhere.

Other readers are trying to assimilate the facts and truth presented with what they already know or believe to be true. At points, one might think 'oh, I know exactly what you mean, I've seen that happen before', or 'that makes no sense of what I already hold to be true, it can't possibly be right, even if it did come from a randomised control trial!' Humans almost always desire a level of coherence of new truths with those they already hold. Coherence is essential because holding opposing sets of beliefs leads to cognitive or moral dissonance.

Throughout, I am providing plenty of practical advice – some of it from research, some of it from lived experience. For many readers, it won't matter much where the advice comes from, as long as it works. These pragmatists will hope for some good tips to avoid betrayal or to recover from a moral injury and might yell, 'Keep your randomised control trial, this works for me!'. And sometimes we blend these schools of thought. Suppose a reader were to combine the correspondence and pragmatist schools. In that case, they might ask, 'Yes, that evidence is statistically significant, but is it clinically significant?', or in other words, there is scientific evidence, but does it also work in the real world?

When I told Mike's story in the introduction or Alice's story in Chapter 3, I presented a personal, constructive interpretation of the events of betrayal. However, if you asked Mike's manager or Alice's former husband, they might construct a different understanding of those events. Telling stories and being sensitive to context are important. It was important to recount the betrayals in Mike's and Alice's stories because the contextual richness of their experiences provides a depth of emotion that is innately human and resonates strongly with many people.

Most people use a combination of all four schools of thought when working out what is true. We quite rightly blend these, and the way we mix the elements forms the deeply held moral expectations that we bring into the workplace. Sometimes the betrayal we feel may be a matter of adjudicating the manner in which we arrived at the truth. A major difficulty arises when someone has more power and can dominate the other person's understanding of events. In telling his account, Clinton did so with much more power and ambiguous language than Lewinsky was able to access to address his misleading statements.

Truth in a Post-Truth World

Truth now exists in what some are calling a 'post-truth' world. This term is possibly misleading, as we have actually moved towards perspectival truths that I have alluded to above as 'my truth'. These personal truths often overlook facts inconvenient to one's position. Telling 'my truth' allows the powerless a platform to speak to and against power structures. However, the void of objective truth has many downsides. As scandals reveal, we have reasonable cause to believe that former and current political leaders may not be entirely truthful about their sexual exploits, shareholdings, or allegiance to corporate sponsors. These leaders can claim these half-truths are their truths. However, objective truth has consequences, not all of which are personally or corporately desirable.

Increasingly, constructivism has allowed and encouraged people to embrace 'my truth' rather than 'the truth'. This sounds fine in many ways, as we each experience reality in different ways, through factors such as gender, race, or social class, for example. The constructivist approach helpfully allows us to take more seriously that truths rising from historical events are experienced differently depending on whether you are victor or vanquished. Both deserve to have their stories heard.

However, a serious downside to 'my truth' is when it is more than embraced, it is enforced on others. With clashes of various people's truths increasing, it seems we also have fewer tools to adjudicate whose pile of bricks more closely resembles a house. This matters in the workplace because potentially morally injurious workplace events often involve a clash of truth claims. When an individual claims to have been morally injured by a workplace policy, they can enforce what some have called a moral imposition of their personal moral code and truth claims onto the workplace (Johnstone & Hutchinson, 2015).

Nursing Professor Megan-Jane Johnstone recognises that when someone enforces their personal truth or moral code in the workplace, it can override the moral frameworks of colleagues and organisations. 'My truth', therefore, morphs from a freedom of expression to a personal moral castle from which to lob grenades at others. This truth warfare extends beyond workplaces; it is a dangerous cultural milieu.

Media outlets and the political class on both ends of the spectrum rally against the lack of truthfulness in public discourse. Those to the left of the political spectrum tend to label factual opacity as misinformation or disinformation – those to the right use 'fake news' to describe the same allegation. What astute observers agree upon is that polemics is increasing, and civility is dying out. We are now lobbing more and more violent moral accusations at those with whom we disagree.

This unwillingness to consider other viewpoints could not have been more eloquently put than by Federal Court Judge, Justice Michael Lee, who presided over a recent Australian defamation trial centring around the claims and counterclaims of an alleged sexual assault in Parliament House. Justice Lee produced a lengthy, articulate, and wise ruling, which opened with these remarks:

> For some people, any unwelcome findings will be peremptorily dismissed. The reasoning process, including the drawing of fine distinctions based upon the subtleties of the evidence, will be of no interest. This reaction is inevitable given that several observers have a Rorschach test-like response to this controversy and fasten doggedly upon the "truth" as they perceive it. Their response is visceral because the "truth" is revealed and declaimed, rather than proven and explained. Some jump to predetermined conclusions because they are disposed to be sceptical about complaints of sexual assault and hold stereotyped beliefs about the expected behaviour of rape victims, described by social scientists as "rape myths"; others say they "believe all women", surrendering their critical faculties by embracing and acting upon a slogan arising out of the #MeToo movement. Some have predetermined views as to the existence or otherwise of a conspiracy to suppress a rape for political purposes. For more than a few, this dispute has become a proxy for broader cultural and political conflicts.

> This judgment is not written for people who have made up their mind before any evidence was adduced or are content to rest upon preconceived opinions. It is written to set out my factual findings comprehensively and explain my decision to the parties and to the open and fair minded.
>
> *The Honourable Michael Lee, 2024*

Careful moral engagement requires a humility to admit no one possesses all the facts, that we may have biased views of those facts, and maintain an openness to other viewpoints or mutual civility (see Chapter 4). Before accusing a leader or organisation of morally injuring us, it is worth making space in 'my truth' to accommodate other viewpoints through 'enlarged thinking' (see Chapter 10).

Summing It Up: The Truth

The sum of the above argument is that truth-telling, not just projecting one's own personal understanding of truth or morality, is vital in non-morally injurious workplaces. The manipulation of truth is central to the breaking of trust, which is the act of betrayal. Truth is not infinitely malleable, and people recognise when it is stretched beyond reasonable bounds. Organisations often operate in the realm of ambiguity by using corporate spin that leverages the space between truth and a lie.

For Clinton, the difference between physical and sexual or between present and past tense was exploited in a way that many sense is troubling. For the ADF, the development of new programmes masked the reality that they did less to help staff than appearances would suggest. For a large organisation or a powerful leader, their truth has a way of prevailing because of the resources to discredit other versions of the truth.

Alternatively, individual employees may have unrealistic expectations about how their own moral code might be applied in their day-to-day work. In any workplace, compromise does not necessitate dissonance. Beware of equating 'my truth' as 'the truth' which we demand everyone must bend around. There are five points to take away from this section.

1. Facts are not truths: the grey area between facts and truth, or between truth and lies, is large, often leveraged to betray workers and managers alike. The act of paltering is used in the gap to produce deniability and lower the chances of being held accountable.
2. Truth matters: without truth, trust in our workplace dies, and the attendant stench makes the workplace destructive to personal and corporate wellbeing. Workplace leaders need to both convey the truth and navigate multiple competing truth claims.
3. Truth is tricky: not everyone sees moral matters the same way. As much as an individual might cry foul over what has happened to them, they must seek to understand other viewpoints before claiming to be morally injured when their truth doesn't rule. Individuals will need the proactive moral self-reflection practices detailed in Part 3.

4. Moral frameworks are built on truth claims: moral dissonance is the clash of competing truth claims as well as moral visions. The telling of untruths via lies and paltering is central to workplace betrayal. The competition of truth claims may manifest as workplace disputes that are harder to resolve than a dispute over pay and conditions.
5. Humans are still suckers for the truth: each of us has trouble telling the whole truth when it is required of us, but we desire to hear the whole truth from others. Integrity demands we love telling the truth as much as we love hearing it.

In conclusion, I argue that for our purposes, truth construction is best understood as a process of humbly and civility arranging our moral beliefs in coherent systems that align with real-world facts and avoids the temptation to palter in our interactions in the workplace.

Truth is the foundation for trust – we feel we can trust people who tell the truth, and truth-telling creates the conditions for trust to flourish. Interestingly, truth and trust share the same etymological root with the word 'tree'. Experientially, the truth-teller resembles the steadfastness of a tree firmly rooted in the ground, which provides the fruit and shade of truth for those who draw near in trust. Lies chop the tree down, while paltering removes one branch at a time, sometimes with lethal effect.

LEVERAGING THE GREY: PALTERING

When the two new Boeing 737 MAX aircraft crashed (see Chapter 4), Boeing initially paltered in its press releases, with some considering their statements to be strategically misleading. Boeing noted it was 'deeply saddened' and emphasised how it was supporting local investigators of the crashes. Boeing made themselves look both sad and helpful, almost as if they were as much a victim as those who lost their lives or loved ones. Boeing simultaneously heeded Gray and Wegner's advice to be both the hero and the victim as a way to minimise blame (see Chapter 5).

Subsequent investigations of these crashes revealed a litany of procedural and cultural issues at Boeing that were paltered. What Boeing's statements avoided acknowledging were the known design flaws in the novel manoeuvring system installed in the new aircraft. The manoeuvring system would eventually be shown as the primary contributor to both crashes. Nor did Boeing representatives mention internal email communications that had belittled one of the airlines involved in the crash for querying minimal pilot training to fly the new aircraft. Boeing did not mention the culture change from engineering excellence to profit-making, but instead emphasised procedural cooperation and emotional expressions to create an impression of responsibility and responsiveness.

Their paltering helped shield Boeing from immediate accountability while maintaining a surface appearance of 'transparency' and concern. Paltering is not simply the domain of corporate press releases – it is used daily in organisational betrayal to avoid truth-telling, to gaslight whistleblowers, and to prevent accusations that the halo might be slipping.

Another person approached me during my PhD about an incident they had experienced in a faith-based not-for-profit organisation. Jenny had worked for the organisation for several years, but some differences in style and direction had arisen within her team. Members of that team complained about Jenny to the director. Jenny admitted there were lessons she needed to learn and improvements she could have made. However, Jenny was never allowed to discuss, let alone act on, those lessons. Instead, secret conversations were had about Jenny by senior leadership before she was summoned to the director's office, and at that meeting, Jenny was told of her termination.

Besides being terminated without due legal process, the director tried to make Jenny complicit in their forthcoming paltering about her termination. Jenny was told that the termination would not be announced publicly until she had found new employment, so that it could be communicated that 'Jenny is leaving to pursue an exciting new opportunity'. When Jenny found that new role, the statement was not strictly a lie, but it was not truthful either – she was both gaslit and paltered in one move.

Paltering is a central technique in the defence phase of DARVO that is hard, if not impossible, to debunk and unlikely to be legally sanctioned. I say this with deep sadness – many altruistic or faith-based organisations I have worked for are among the worst at these kinds of cover-ups. At one level, it makes perfect sense. They have held a revered name and have so much to lose if their actions are seen for what they are. But it is doubly disturbing that such moral dissonance is possible in supposedly upright community leaders. Due to its potential to harm, the development of a paltering detection system is crucial.

Linguistic Contortions

I have a deep admiration for Olympic gymnasts. They are likely the strongest and most skilled athletes at the Games. One cannot help but be amazed at their ability to manoeuvre their bodies around a floor, pommel horse, or parallel bars. Similarly, I deeply respect a wordsmith who can manoeuvre language in a poem, novel, or paper to express truths in a way I would not have previously considered. Magician's assistants, much like gymnasts, can do amazing things with their bodies. To fool their audience, they can fold their body in half, creating the illusion that they have been sawn in half. Paltering is not the skilled manoeuvring of a gymnast or a poet; it is the skilled misdirection of a magician's assistant – a mere linguistic trickster.

Schauer and Zeckhauser describe paltering as the practice of fudging, twisting, shading, bending, stretching, slanting, exaggerating, distorting, whitewashing, and selective reporting (Schauer & Zeckhauser, 2007). Sociology of religion Professor Linda Woodhead MBE describes paltering as the art of blanking, silencing, spinning, overloading with information, conducting lengthy investigations, writing reports that are then buried, involving lawyers, invoking confidentiality, and issuing non-disclosure settlements (Woodhead, 2022).

These acts of verbal contortion, which are the heart and soul of paltering, are dangerous because they enable the misuse of legal and administrative processes. The detection system requires examining fine-tuned linguistic usages that differentiate between gymnasts and magicians, as well as between poets and con artists.

Fine Distinctions

In the 1980s many English-speaking countries abandoned the study of grammar. As a result, an entire generation, including mine, did not learn how language works. Understanding the construction of a prepositional clause or a split infinitive was beyond me, much to my mother's chagrin. I only learnt about grammar when I studied a foreign language in my thirties, and in understanding how language works, I was enabled to perceive how paltering is committed.

Palterers work grammar to their advantage by the use of highly specific linguistic constructions, allowing them to deny their misleading ambiguity or doublespeak later. Grammar contortions were a tactic employed by Clinton, such as the use of tense and stretching the semantic limits of words. The use of one or two verbal twists is not a definitive sign of paltering; however, recurring patterns or an accumulation of red flag language use will alert the discerning observer that the whole truth is being avoided.

A subtle sign of paltering is the use of the passive voice. Passive voice changes a verb from an act one does to an action being done to the speaker. For example, instead of admitting 'we made mistakes', errors are expressed as 'mistakes were made'. For an incident involving injuries or loss, a palterer will regret injuries 'caused to people' rather than 'injuries resulting from a crash'. Internal communications work similarly. If you raise a concern in the workplace and hear 'your concern has been acknowledged…' rather than 'these are the steps we will take to address this…', it is possible a palter is occurring.

Alongside the passivity, paltering will often omit who is acknowledging the facts, or who made mistakes. Admittedly, naming who made mistakes in the opening stages of any investigation is legally fraught and highly susceptible to abuse (see Chapter 7). However, a point arrives when truthfulness demands that a person, role, or process be directly named. Truth demands that an organisation or leader clearly state who will or did act. Truth is precise, paltering is vague.

Another key indicator of paltering is when an organisation or leader praises their own actions well before they have done anything praiseworthy. When something like a safety concern or data leakage has occurred, they may say 'we proactively notified customers'. Of course they notified customers; it was most likely a legal obligation. However, it is possible to cast oneself as the hero by simply saying one did what they were compelled to do. These kinds of statements are paltering because they fail to admit the action was an obligation, rather than a chosen option. When a concern in the workplace is raised, a leader may say, 'Matters like this are taken seriously'. It is unclear who takes it seriously and whether they are taking the issue or your audacity to raise the complaint seriously. Regardless, they have painted themselves in a positive and ambiguous light.

Another aspect of paltering is the expression of sadness or sympathy for those affected. Many executives are, in fact, saddened by adverse events; however, when expressions of corporate sympathy are prioritised over active empathy for those impacted, we have reasonable cause to be suspicious. For example, in 2010, former

UK Prime Minister Gordon Brown was recorded making disparaging comments about a constituent. In what he thought was the privacy of his chauffeur-driven car, Brown said, ‘She was just a sort of bigoted woman’. When this was released, his first reaction was mortification: ‘I apologise if I’ve said anything that has been hurtful’. Many of the elements of paltering were present – passive voice (has been hurtful), vagueness (if I’ve said) and his own sadness. The classic paltering move is to say, ‘I’m sorry that you have been hurt’. This statement misdirects, implying that the fault lies with the person who dares to be hurt. Now, if I hear this, I bluntly reply, ‘Don’t apologise for my reaction, apologise for your action’.

Brown’s fudging of personal mortification, misdirecting blame, and his expression of sympathy, rather than empathy, did not work in the public’s mind. To his credit, Brown later said directly, ‘I was in the wrong’, but still felt the need to modify this by saying the wrong was his misunderstanding and misjudgement rather than the harsh words. In the workplace, palters come in the form of insincere phrases such as ‘I am sorry you feel that way’, ‘it wasn’t my intention for that to happen’, and ‘it was a misunderstanding’.

Take seriously what is being said when you feel you may be in the defence and attack phases of betrayal. Where possible, seek clarification of any vagueness, misdirection, passive verb usage, seemingly innocuous use of tense, and use of pronouns that mark out hubris. When these linguistic contortions are left unchallenged, their ambiguity will be the saviour of the person betraying you. Clarity of grammar is your friend, and so is being alert to the presence of weasel words.

Weasel Words

The term ‘weasel word’ was first coined by author Stewart Chaplin in a piece called ‘The stained-glass political platform’. Chaplin wrote,

> Weasel words are words that suck all the life out of the words next to them, just as a weasel sucks an egg and leaves the shell. If you heft the egg afterward it’s as light as a feather, and not very filling when you’re hungry; but a basketful of them would make quite a show, and would bamboozle the unwary.
>
> Chaplin, 1900

Weasel words are designed to look impressive but are in fact words utilised to bamboozle the listener. Weasel words might include ‘unprecedented’ in place of ‘what we were warned might happen’, the use of ‘anomaly’ instead of ‘failure’ or ‘defect’, or the use of ‘adverse outcome’ instead of ‘crash’ or ‘explosion’. Someone in the publicity department might suggest saying, ‘experts in the field’ rather than ‘our in-house manager who has an enormous conflict of interest’, will conduct an ‘ongoing investigation’ in place of ‘working out who we credibly can blame for this’, and ‘regrettable loss’ rather than ‘death’. A weasel-worded statement about a factory explosion might look like this:

> Over the course of the last evening, an unprecedented technical anomaly produced a notable adverse outcome at the manufacturing plant that has provided jobs for hundreds of local workers. We are currently having a range

> of experts assist in what will be an ongoing investigation into the anomaly. The regrettable loss of members of our valued workforce deeply saddens us.

A more truthful way of stating it would be:

> Last night, an explosion destroyed our factory and killed 12 of the lowest-paid workers in the community. We were warned this explosion might happen, but fixing it was going to be expensive. The off-duty manager who ignored the warning is currently working out who else can be blamed for these preventable deaths so that our share price can rebound in a couple of weeks.

This is a harsh but fair comparison. Early corporate statements are not about truth but about spin and minimising impact on share prices and protecting those in power. The spin is not borne merely of the moment around critical incidents; it has become a corporate way of life. Weasel words like 'transparency' and 'openness' replace words like 'truth'. Linda Woodhead notes that almost none of those acronym-centric corporate ethics statements contain the word 'truth' (Woodhead, 2022). Organisations vaunt terms like 'accountability' and 'professionalism' and soften the word 'punishment', which is applied unflinchingly to those lower in the chain of command and sparingly to those with more power, as many Sub-postmasters discovered.

Root and branch, many large institutions have a cultural and practical aversion to truth-telling. This aversion is the heart of betrayal. When truths about difficult topics are raised, many organisations and institutions find ways to overlook them and sideline those who raise them. When the truth is required from an organisation, they often find ways to palter.

HOLDING THE LINE: TRUTH AND TRUST

The above analysis may have created the impression that all corporate communication is suspicious. Statements of regret, carefully crafted legal statements, and the like are as suspicious as they are judicious. Indeed, the historical accumulation of corporate paltering has led many to distrust corporate spin automatically. The Edelman Trust Barometer in 2017 reported that 64% of people trust leaked information more than what corporations say in press statements. The truth has not been told in the past, and trust in leadership is increasingly rare for staff and the public.

The antidote is not just to be truthful, but to engage in activities that overtly and genuinely rebuild trust. Rebuilding trust now requires more than words, which, like truth, have become so malleable; it requires action. Organisational leaders must understand, value, and fulfil the covenants they leveraged at recruitment. Leaders must relinquish the habit of giving commands and collaborate with those who are rightly sceptical of empty statements about openness and accountability.

Action, and not words, is required. The 2019 Edelman Trust Barometer reported that 78% of employees believe how a company treats its employees is one of the best indicators of its trustworthiness (Edelman Trust Institute, 2019). In Part 3, we will explore what trust building looks like, but for now turn to scapegoating, the last step in organisational betrayal.

ITEMS FOR CONSIDERATION

- Examine some recent public declarations your organisation has made about 'adverse incidents'. How close are these to being true, and how much are they a palter?
- What blockages do you have in telling the truth rather than paltering? How might you be the change in truth-telling that you would like to see in your organisation?

7 Landing Betrayal
The Art of Scapegoating

A true word.

> *You know nothing at all! You do not realise that it is better for you that one man dies for the people than that the whole nation perish.*
>
> High Priest Caiaphas

> *The search for a scapegoat is the easiest of all hunting expeditions.*
>
> Dwight D. Eisenhower

INTRODUCTION: SYSTEMS FAILURE

In 2014, around the time Monica Lewinsky was reflecting on 16 years of public shame, an arm of the Australian Government started devising a cunning scheme to recoup billions of dollars of losses from those who had received government benefits illegally. The tactic of clamping down on 'welfare cheats' is often lucrative for governments and almost always popular with voters. The Department of Human Services (DHS) first produced a note about a benign-sounding 'Trusted Data Assessments' in 2014. This eventually led to the introduction of a government programme in 2016, which would have three iterations to recoup debts, all with bureaucratically dry acronyms that Orwell would be proud of – but which would eventually be known as *Robodebt* – an illegal and morally dangerous program that would become the subject of yet another Royal Commission.

Some years after the creation of the scheme, Renée Leon was appointed as the Secretary of the DHS that had established the fraught programme.[1] Leon had a prestigious background, graduating from her law class with first-class honours, holding a master's degree from Cambridge, and receiving a national medal for public service. When she moved to serve as Secretary of the DHS, she knew it would be a challenge to take over a department riddled with cultural issues, but she would not have expected to have her career entangled with the failings of *Robodebt*.

DOI: 10.1201/9781003499725-9

A later Royal Commission into *Robodebt* clearly stated that Leon had not been involved in the development and implementation of the scheme. Indeed, it was after mounting public outrage that Leon shut down the scheme in June 2020. Her job was terminated two weeks after her directive to shut down the scheme, ending 30 years of dedicated government service. Despite thousands of public servants being involved in the scheme, she was one of only two public servants to be publicly named by the Australian Public Commissioner. Leon was a scapegoat.

This is not to say that with the benefit of hindsight, Leon performed all her duties without error – with such a complex portfolio and so many regulations, it was inevitable that she would err at some point. However, the Royal Commission into the Robodebt scheme reported that, 'there is no suggestion that Ms Leon was acting in bad faith at any time'. The Royal Commission neither referred her to the Australian Public Service Commission (which it could not legally do), nor to any anti-corruption commission, nor recommended any criminal proceedings against her.

However, she was investigated by the Public Service Commission, despite it having no jurisdiction. Legal affairs editor at The Australian newspaper, Chris Merritt, reported the lengths to which government sources stooped to have her named and shamed. When Leon was referred to the Public Service Commission, there was no legal mandate to do so. Merritt reports that, for more than a year, an investigation proceeded against Leon without a legal basis. To remedy the situation of an unlawful investigation, 11 days before Leon was publicly named, the government introduced retrospective legislation to render what was previously unlawful suddenly lawful. No political leader who approved and promoted the scheme has ever faced sanction.

To summarise, Leon did not start the scheme and, in fact, ended it. She was removed from her role, investigated by a Royal Commission and found to have exercised good faith. She was illegally investigated again, and only through the enactment of retrospective legislation was she named as a perpetrator. Of all the politicians and public servants involved, she became one of two public servants named. If it looks like a scapegoat, sounds like a scapegoat, and smells like a scapegoat, it probably is.

The final action of betrayal mechanisms is to attach blame, excuses, and shame upon a particular person or group. Betrayal may include sacking someone for raising a complaint, but that is illegal, something noted earlier with the sacking of a safety worker by Qantas. Betrayal often goes further by finding ways to hold whistleblowers accountable for some related or unrelated oversight.

The methodology of DARVO requires that a whistleblower's claim not only be denied, but that they also become the problem to be removed. Harking back to Mike's story from the introduction, his attempts to defend vulnerable staff were turned around to accuse him of wrongdoing. The experience of betrayal in the workplace often involves the use of a DARVO process, which begins with halo protection, is discussed through paltering, and culminates in scapegoating. DARVO, halo protection, paltering, and scapegoating are difficult to prove, so they often escape legal sanction; however, collectively, they are morally destructive.

Scapegoating in History and Its Current Practice

In History and the Workplace

The concept of scapegoating originates in the Old Testament of the Bible, specifically in the Book of Leviticus (16:10). It details that on the Day of Atonement (Yom Kippur), two goats were to be chosen: one is sacrificed to God, and the other, the 'scapegoat' is symbolically laden with the sins of the people and then driven into the wilderness. This ritual provided a symbolic mechanism for communal purification and the transfer of guilt. The term 'scapegoat' itself comes from William Tyndale's 1530 English translation of the Bible, where he rendered the Hebrew term *azazel* as 'escape goat', later simplified to 'scapegoat'.

Over time, the term evolved from a theological concept to a broader sociological and psychological one. In modern thought, scapegoating refers to the process by which individuals or groups assign blame to an innocent party to relieve collective guilt and avoid personal responsibility. What began as a religious metaphor and ceremony to cleanse communal guilt has evolved into a central means for societies to assign guilt to one person or group for in-group protection.

Chapter 5 briefly mentioned the work of Dr Nathalie Martinek, who researches scapegoating in modern workplaces. Martinek details how this social mechanism operates in workplaces, so that specific individuals are burdened with blame and exclusion. As Martinek explains, 'The scapegoat serves a sacrificial function to assure group cohesion' through the exclusion of unwanted change, complaint, or concern (Martinek, 2025b).

In narcissistic or dysfunctional organisations, leaders who 'prioritise image, control, and loyalty over reflection, transparency, and repair' (Martinek, 2025a) facilitate scapegoating in a predictable sequence beginning with idealisation. At this initial point, the target is welcomed, praised, and trusted because they appear to embody the group's values. But this early admiration is conditional, and when the newcomer's insight, confidence, or dissent begins to unsettle the group, the tone of communication and work allocation shifts.

Martinek describes a quiet but powerful breakdown: devaluation, humiliation, and narrative control. The person is subtly undermined, socially excluded, and framed as the problem (think DARVO). Suddenly, former allies withdraw as a means of self-protection. Over time, the scapegoat is quietly discarded, often without any formal explanation. This is something Jenny, from Chapter 6, experienced. Finally, scapegoats are either quietly fired or hand in a disillusioned resignation, both events marked by a generic farewell email.

This scapegoating process, Martinek argues, is less about individual conflict than it is about systemic fragility. Healthy systems can tolerate discomfort and difference, while dysfunctional, self-conscious ones cannot. Martinek calls on those who have been scapegoated to move beyond victimhood by recognising the pattern and reclaiming their own agency. She writes, 'The story changes when you stop performing the part you were given and begin telling the truth of what you saw and experienced' (Martinek, 2025a). Some tactics to rewrite your story will be given in Chapter 9. As the case studies below demonstrate, the need for truth-telling extends across various workplaces and organisations.

In Organisations

Public Service

The chapter started with Renée Leon's tale, a former senior public servant around whose neck the *Robodebt* scandal was hung. Leon was reported to have breached 13 different public service standards; however, breaching this many rules seems like a regular day in the public service. Public service in most jurisdictions is undertaken within large systems that have often evolved into *institutions* or *museums* (see Chapter 4). These large entities are almost defined by the vast number of regulations, processes, and procedures that must be complied with. In fact, procedures and regulations have reached such a point that no person can comply with them all. The point at which an administrative breach is detected is enough for betrayal through scapegoating to commence.

When the procedural framework of an organisation becomes so extensive that breaching some rule is unavoidable, those who work within the system have become the enemy. An accumulation of rules can serve to diminish fair and equitable treatment of those in the organisation. Cicero reportedly said, 'The more laws, the less justice', words resonating with today's institutional regulation maze. Work Health and Safety expert Greg Smith reports a legal finding that supports Cicero's contention: 'site rules and regulations could translate into arbitrary and punitive forms of punishment, which undermined participants' ability to fulfil jobs to their satisfaction and left them feeling insecure with their positions' (Sellenger Centre for Research in Law, Justice and Social Change as reported in Smith, 2018).

Furthermore, when an institution or museum has numerous rules, it often has considerable resources that a dedicated staff member, if not an entire legal department, can utilise to identify breaches that a whistle-blower has committed. I met Amanda some years ago, a senior executive who was progressing well in her career but was also causing some waves about the culture and direction of the organisation. A series of complaints was lodged against Amanda, with two of them regarding possible breaches of procurement processes. Investigations were launched but stalled for many months.

Over those months, such psychological pressure mounted on Amanda that she was in significant distress and forced to hire lawyers at her own expense to bring a resolution. A week after engaging lawyers, she was informed that the stalled report had been concluded, and she had no case to answer. The irony is that she held a key leadership role in procurement oversight that was never halted or removed. The tragedy was that no remediation or apology came regarding the months of stress, while communication about her matter was non-existent. It is hard not to see this as a form of administrative violence and a failed attempt at scapegoating. While the attempt was unsuccessful, Amanda was put in her place, and I believe morally injured.

In public service, the use of scapegoating often involves a considerable amount of administrative violence related to compliance with overly bureaucratic regimes. Areas that can be leveraged for scapegoating are the over-regulated fields of procurement, work health and safety, and the ever-increasing vagaries of workplace respect. Vexatious organisational investigations can also involve non-compliance with the weasel-worded 'transparency', 'openness', and 'accountability'. Be warned that if you work in the public service, the systems that are there to protect you may also be used to scapegoat you.

Not for Profits

Let's return to Mike's story of betrayal from the introduction. Mike worked in a not-for-profit that proudly touted its good work in the community and its inclusive, forward-thinking mindset, developing and nurturing a public-facing reputation for doing good and being good. Its annual revenues were over one billion Australian dollars, but many of its employees were among the worst-paid in the country. This not-for-profit was an institution with numerous senior executives, numerous processes, a flashy values statement (excluding truth), and departments for legal, PR, HR, and WHS. It appeared to be a charity but functioned like a corporation. Mike had upset the wrong people, so DARVO was deployed against him.

A couple of months before the deployment of DARVO against Mike, he received over a dozen complaints that were verifiable by the emails a senior manager frequently sent to all staff. The emails were regularly in large font, capitalised, and written in red for full effect. The communications reminded staff they were under constant video monitoring, footage from which would be used to justify the confiscation of personal property or remove basic working conditions. Additionally, these emails regularly included the public naming and shaming of individual staff members. The poison here was in the very structure of the management and culture of the worksite and was clearly distressing.

The main reason given for not acting against the abusive manager was that, in a highly regulated industry, they met more compliance targets than most managers. Altruistic institutions often hold their values in high regard publicly but regularly lose sight of them, to prioritise their reputation while truthfulness and compassion are effectively abandoned. If you are in a not-for-profit, do not believe, even for a moment, that it will not act out of self-defence and attempt to scapegoat you. In fact, they have greater reason to do so than other organisations.

In a not-for-profit, the process begins with whispered conversations in hallways that are strained and awkward. At first, a statement is made that any alleged poor conduct 'is not who we are', followed by a paltered passive declaration that something should be done. But when something starts to get done, silence enters where dialogue once existed. After rumours of concern have circulated widely, possibly to senior management, allegations will arise against whistleblowers for unrelated issues. Very little will be public from the time of the allegation until the removal of the whistleblower and the inevitable group email: 'Mike has moved on to pursue exciting opportunities'.

Most of this process will be done off the books so that no real trail can be found. As few records will be kept, it is essential to ensure that you make notes and confirm any directions you are given from managers via email. Don't forget to record non-replies. If feasible, keep those emails to protect yourself from future allegations. Be careful of sharing emails with outside agencies, however, as some not-for-profits are more likely to pursue you for breaching data protection laws than they are to pursue abuse by managers.

Industry

Betrayal in industry does not have the immediate incongruence it does in an altruistic not-for-profit – in some ways, it is part of doing business. To understand the business model of betrayal, let's revisit the Boeing 737 MAX scandal. Businesses will still DARVO, palter, and scapegoat so-called troublemakers. One interesting tactic used as a precursor to betrayal is for a company to seek to remove itself from any regulatory oversight.

From the beginning of the 737 MAX's design, Boeing advocated for increased delegation of engineering compliance away from US Federal Aviation Administration (FAA) oversight, allowing compliance to be performed internally. A 2021 Department of Transportation Inspector General report found that while the FAA followed certification procedures, these relied too much on Boeing's own interpretations. In-house assessments resulted in significant misrepresentations of the manoeuvring system that largely led to the crashes.

Now, recall also that the shift in headquarters from Seattle to Chicago was a deliberate cultural change aimed at increasing profit. Both *The Washington Post* and *The New York Times* reported that Boeing's efforts to convince the FAA to relax certification requirements, including crew-training provisions, aimed to expedite delivery, remain competitive with Airbus, and increase profitability. The drive for profit led to a decrease in accountability to external agencies.

Much was made of Boeing's deleterious culture at the time, and CEO Dennis Muilenburg was held accountable. However, this case highlights how 'accountability' has become a weasel word. It is arguable whether Muilenburg's departure was a termination, resignation, or a board-pressured resignation. This act of 'accountability' resulted in Muilenburg walking away with pay, stocks, and pensions worth USD $62 million. Many of us would like to be held accountable in this manner. However, another man had a different kind of accountability.

As of the time of writing, only one Boeing employee has faced criminal charges directly related to the 737 MAX crashes. Former Chief Technical Pilot Mark Forkner was held accountable, but was in many ways a scapegoat. In October 2021, he was indicted on charges of fraud for allegedly misleading the FAA about the Manoeuvring Characteristics Augmentation System (MCAS) flight control system. Forkner was accused of providing false information that led to the omission of MCAS details from pilot training materials. In March 2022, a jury acquitted him of all charges. No other Boeing employee has been criminally or civilly charged in connection with the 737 MAX incidents despite the knowing involvement of many executives.

What this saga teaches us is that 'accountability' in many industries is a farce. Boeing sought to avoid regulatory accountability, allowed its CEO to leave with a fortune, while one lone person faced prosecution. In the industry, many leaders will walk away into a new, high-paying role, while those in the middle face the brunt of the blowtorch. This is the lesson that Renée Leon and Mark Forkner both learnt the hard way.

Therefore, be aware that if you are currently ensconced in upper middle management, you may become a scapegoat for organisational failings. Middle managers inhabit the peculiarly vulnerable position of being significant enough to display that the organisation is taking any scandal they face 'seriously' by removing them. Middle managers are also low enough in the hierarchy not to get large pay-outs or a large legal team to defend their interests. Middle managers are also often overworked, so at some point, they have made an oversight for which they can be 'held accountable'. Perhaps middle managers are worthy of more sympathy from senior executives and frontline workers.

A Foundation for Scapegoating: Bad Apples

When Enron, a large US energy services company, collapsed in 2001, the fallout exposed one of the most elaborate accounting frauds in US corporate history. The saga demonstrated the ease with which an organisation can attribute complex wrongdoing to an individual. The public was provided with a scapegoat.

While multiple executives were involved in the company's deceptive financial practices, Andrew Fastow, the company's Chief Financial Officer, was quickly positioned as the central villain. Media outlets and prosecutors cast him as the mastermind of Enron's complex off-the-record partnerships, conveniently centring the scandal around him. As *The New Yorker* observed, Fastow became the main focal point of the investigation, serving as the scapegoat that absorbed the weight of public outrage and institutional failure.

However, investigations, including the US Senate's Permanent Subcommittee on Investigations report, revealed that Fastow operated within a permissive system marked by conflicted auditors, disengaged board oversight, and a corporate culture that rewarded aggressive financial manipulation. Still, the narrative of one bad apple persisted. The inference was that the organisation itself was sound. However, as the story was told, Enron had been corrupted by one bad apple. Bad apple theory is the focus of attention by Safety Science Professor, 737 pilot, chaplain, and Professor Sidney Dekker. He writes: 'When we say the cause of an accident is a bad apple, we imply that the system works well—except for that one individual' (Dekker, 2014).

Dekker critiques the bad apple theory as 'intellectually lazy and morally bankrupt' (Dekker, 2007). It offers a tidy sense of moral and intellectual closure, especially to senior leaders and oversight bodies who may have failed in their duties. By isolating blame to one person alone, attention is deflected from structural issues such as excessive numbers or complexity of processes, poor checks and balances, and permissive leadership. The bad apple becomes the sacrificial figure through whom an organisation's betrayal of its own values, let alone its workers, is expiated.

When viewed together, the bad apple theory and scapegoating are a marriage made in heaven. The pairing avoids organisation-wide accountability, or accountability for the C-suite complicit in the creation and curation of an environment that empowered the supposed wrongdoer. A person who is going through the system of wearing blame is isolated and vulnerable because they are blamed by management, avoided by colleagues and feel the weight of betrayal. There is an urgent need to reconsider whether bad apples or bad barrels – broken workers or broken workplaces – require the most accountability.

Bad Apples or Bad Systems?

Dekker and others advocate for a systems approach to error and misconduct that seeks to understand the broader context in which harmful decisions are made. This perspective does not absolve individuals of responsibility but resists simplistic narratives of individual culpability. It asks not just who failed, but how the system made failure more likely. We must, therefore, consider the environment in which Mike's manager, Fastow, Forkner, Lewinsky, Leon, and many others worked.

As Dekker explains, human error is not the cause of failure but a symptom of deeper trouble within a system. Latent conditions, such as unclear policies, poor communication channels, or unmanageable workloads, can silently shape behaviour long before visible failure occurs. Yet many organisational leaders resist systems thinking, particularly when wrongdoing or harm has happened for which they may be held accountable.

A systems approach often implies that leadership, not just frontline actors, played a role in enabling misconduct. This can trigger deep institutional defensiveness and the inevitable creation of scapegoats. Acknowledging systemic factors may expose design flaws, cultural dysfunction, or neglected warnings, all of which may reflect poorly on those in charge. As a result, organisations often find it easier to default to bad apple narratives and scapegoating tactics. It localises blame, protects reputations, and allows broken or toxic systems to continue undisturbed.

For those caught in the machinery of scapegoating, the experience can be deeply wounding. When a workplace refuses to confront the systemic nature of its failures, it not only scapegoats individuals – it inflicts moral injury on those who see through the façade.

In the long term, if and when these tactics are revealed, it can lead to a deep loss of institutional credibility, enormous compensation payments, and possibly organisational collapse. These negative impacts, as the annual Edelman Trust Barometer continues to report in increasingly dire terms, are infecting the entire world. A sense of grievance and a desire for vengeance are moving from the radical fringes into mainstream culture. The time for organisations to act is here. In Part 3, an action plan will be laid out so we can reclaim control of this organisational and cultural death spiral.

ITEMS FOR CONSIDERATION

- Public service, not-for-profits, and industry use different elements of their context to betray their staff – excessive rules, feigned concern, or regulatory detachment. Have you seen this in your organisation, and if so, how might it be addressed?
- Middle managers live precarious lives. How can they protect themselves from being scapegoats?
- Think of the last workplace failure you know of. What system errors were in play?

NOTE

1 It must be noted for the sake of transparency/openness, dare I say truth, that Renée Leon is the Vice-Chancellor at the University from which I received my PhD, and am now a sessional academic and researcher. I have sent one message to Leon, which was not replied to and have no other direct contact with her. I was alerted to her predicament by an article by Chris Merritt, in the Australian Newspaper. Merrit is a colleague of my wife. Make of this what you will.

Part 3

A Better Narrative

8 Et Tu and Mea Culpa

A true word.

> *Truth will come to light; murder cannot be hid long; a man's son may, but in the end truth will out.*
>
> William Shakespeare: The Merchant of Venice

INTRODUCTION: UNDER PRESSURE

I have distinct memories of meal preparations in my childhood home. I used to perch myself at the kitchen bench, hoping for some tasty early samples. One strong memory is my mother's preferred method for cooking vegetables – the pressure cooker. Using a sealed pot with a pressure discharge valve, the right amount of pressure can be built up to cook vegetables quickly. On top of the pressure release valve was a small weight that sealed the valve until, suddenly and quite dramatically, the pressure exceeded the stopper's weight. A loud jet of steam made us jump with alarm as it violently shot hot vapour across the kitchen.

Humans can be a little like pressure cookers – we, too, can only take so much pressure until we vent dramatically. Think of the 'sudden' marriage walkout after 20 years of one partner's continual selfishness. Perhaps you have witnessed the office yelling match after a fifth year of 'one-off' cuts to staffing levels. People and systems take pressure… until they don't. When systems can no longer absorb pressure, it's obvious and sometimes fatal.

A violent release of pressure is one way of looking at Brutus' stabbing of Julius Caesar. Brutus is immortalised as a betrayer by Caesar's last words – *et tu Brute*? But was the betrayal merely one way? Was Caesar's slow shift towards dictatorship more pressure than Brutus could bear? Did Brutus feel betrayed by Caesar before his act of desperate violence? We might ask the same of those who increasingly seek to justify grievance-fuelled, hostile activism. I believe understanding the perceived provocations for violence is vitally helpful for understanding the dynamics of violence, but provocation is not an excuse for violence. I always hope leaders do not face violence. I hope, instead, that if they oversee systems that condone betrayal and inflict moral injuries, they deserve a public *mea culpa*.

DOI: 10.1201/9781003499725-11

MEA CULPA

Mea culpa is a Latin term that strictly means 'through my fault', or less formally, 'my fault'. It originated in Roman Catholic liturgy as part of a prayer of confession, but now describes a moment when one admits one's errors or wrongdoings publicly. I was alerted to the importance of *mea culpas* in the revealing of organisational betrayal by a former police officer and soldier, Haydn McComas. McComas was speaking about his Churchill Fellowship at a briefing he gave to members of the Australasian Institute of Emergency Services.

McComas defines a *mea culpa* as the public revelation of misconduct that not only damages an agency's internal cohesion but also threatens its external reputation and social licence to operate. In nature, McComas adds, they are 'usually unexpected, but always explosive' much like the pressure cooker (McComas, 2023). In his fellowship report, McComas looked at leadership and organisational malfeasance in volunteer fire services. Leaders and organisations, he found, covered up these practices until they were forced to admit them in a moment of public contrition.

McComas found that *mea culpas* only occurred when the pressure to admit error exceeded the ability to admit organisational cover-ups. The *mea culpa* is a pivotal crisis in which longstanding organisational malfeasance is finally exposed to public scrutiny. These crises, often portrayed as sudden, are in fact the culmination of a prolonged drift from the stated moral compass of the organisation, resulting from the neglect of leadership as a discipline that transcends mere management of assets.

> Fire services are universally loved… Yet, multiple events of bullying, hazing, sexual harassment, and/or sexual assault in fire services exist, sometimes festering over many years. The common causative factors [are] no leadership, learning investment, and a failure of leaders at all levels to stringently enforce standards to the point that the agency's social licence may be voided.
>
> *McComas, 2023, p. 9*

McComas details several Australian examples where such moments triggered institutional introspection and reform. In 2020, an internal culture survey revealed bullying and sexual harassment targeting women and LGBTI volunteers in the Victorian State Emergency Service Volunteers Association. In 2021, the NSW Rural Fire Service faced significant allegations, leading to an independent Grievance and Discipline Survey by Clayton Utz. Deep-seated issues of mistreatment and a marked lack of leadership capability were revealed. In 2022, an external review exposed widespread cultural problems, including bullying, hazing, and sexual harassment in the Victorian Country Fire Authority.

The resultant damage to these organisations' social contracts no doubt feeds into the steady decline in the number of volunteer firefighters across the country. The Australian Productivity Commission now considers that lower volunteer firefighting capacity poses a significant national risk to Australia. Unless these crises can be circumvented, the personal, organisational, and societal costs will continue to climb.

Mea culpas for these fire services were not inevitable but arose because of a recalcitrance in senior managers who failed to acknowledge or address underlying cultural issues. When people inside the organisation spoke out, instead of being listened to, they were paltered and scapegoated out of the service. McComas writes that these are systemic leadership and governance failures, only rectified through cultural humility, structural change, and a renewed investment in ethical leadership. He frames this not just as an organisational imperative but a moral one, 'To ignore this critical need would be beyond false economy – it would represent the ultimate leadership failure' (McComas, 2023, p. 20).

This chapter aims to provide readers with the tools to build the pressure required for organisational leaders to moments of introspection or face the ritual shaming of a *mea culpa*. When one can sense their own increasing moral dissonance at the culture and practices of their workplace, the pressure is no doubt building because it is unlikely that you are 'the only one' who feels it. Accordingly, we will suggest tactics that may prompt reflection or, if unheeded, a *mea culpa*. These methods are not guaranteed to prevent blowback on those using them. In fact, if you are right about the culture, they will likely provoke strong resistance. So… buyer beware. However, some tactics may protect a whistleblower from being left without support when the paltering and scapegoating begin.

This book is not hoping to tear down organisations and their leaders. The aim is reformation, not revolution. Therefore, four organisational mind shifts are discussed, which may reduce the pressure leading to a *mea culpa*. The hope is that organisations will reduce the risk of workplace betrayals and moral injury, and prevent *mea culpa* moments through meaningful engagement between staff and leaders. Finally, the costs incurred by organisations that fail to act will be outlined as a final plea for organisations to take action.

Inducing a Mea Culpa Moment: A Guide for Employees

If you work in an organisation with many poisonous traits that you are seeing with greater clarity because of this book, you have a choice of three options. First, do nothing and keep grinding away in a place you hate, continuing to take your frustration out on family and loved ones. This is a commonly chosen option, but I don't recommend it for your own health and wellbeing or the stability of your personal relationships.

Second, you can find another place to work. Leaving a place that palters and scapegoats people before it happens to you is a very, very good option. I've walked away from tenured roles for that very reason, so this is not a bad option – it could be very wise. Plan and execute your departure with discretion, avoiding leaving without another role to go to. Third, you can work to change the culture or induce a *mea culpa* moment.

Count the Cost First

If you are considering option three, beware – there be dragons there! If you have seen halo protection, paltering, and scapegoating in your workplace when people raise concerns, by logical necessity, it will happen to you too. Therefore, Mike from the introduction wants you to make a sober judgement about whether you are willing to pay the cost of what you are about to do. Could you financially afford to be unjustly sanctioned and sacked? Before you embark on this road, read, mark, and inwardly digest the advice below lest you are badly burnt.

Also, count the cost on your sense of self and meaning-making if you do nothing. Doing nothing may cost you your soul; doing something may cost you your job. The most expensive thing you have is your integrity. If you have chosen the 'do nothing' or 'leave' option, please read on – it may help you identify individuals who need encouragement as they navigate the poisonous environment in your workplace.

Friends and Relations

Being scapegoated at work is demoralising and socially isolating. Being targeted may entice you to direct your frustrations at those who love and support you the most, until the pressure cooker vents at home. Therefore, before seeking to induce a *mea culpa*, you must have full support from your family and loved ones, as it will be almost as tough on them as on you. If you are unable to verbalise and strategise with them, you may make a plan to address workplace poison that, unlike your family, is blind to your own weaknesses.

Your family is a key provider of the self-knowledge you may well lack; I know they are for me. Part 1 started with the ancient advice to 'know thyself', and Chapter 9 will give you tactics to grow in self-knowledge. However, your family and friends can help you determine if this battle is one you should fight. They may think it will destroy you, or that it fits a personality trait in which you tackle minor problems with big solutions. Their advice may hurt, but not as much as being scapegoated for taking on a trivial issue. Recognise that your family may be the only ones who will give you an empathetic space to recover.

I also recommend finding a mentor to offer guidance and support on your plans. Mike chose to confide in a manager at work, who was compromised by the abusive manager. Many of those in your workplace who utter, 'something should be done, and I will support you', will buckle under pressure and leave you isolated. A mentor should be external to your workplace and someone you have known for a long time, trusting them implicitly. I recommend someone ten years older or more – they may have valuable experience to offer.

A Trail of Crumbs

Create your own records and document the pattern of misconduct or systemic issues by keeping emails, meeting notes, and HR complaints. Also, please note that not all database records will be permanent or accessible. In a recent UK whistleblower case, a senior executive deleted 90,000 emails from the system at the very time a colleague was giving evidence in court about the matter. Researchers on whistleblowing note that 'Systematic documentation is critical… institutional denial often collapses when patterns emerge' (Near & Miceli, 1996). Therefore, you must keep good records… carefully.

Keeping records will be essential to determine whether the behaviour of concern is a single incident or systemic, as your experience of the organisation may not be representative of the organisation as a whole. If you discover that it is widespread, your case for reform is stronger. For example, Mike showed me a note he had made at the time, listing ten names of people who approached him and the substance of their complaints about the abusive manager. If you cannot find any other cases of misconduct, consider seeking advice from your mentor, trusted friends, and family.

When keeping records, be aware that your work computer will be monitored for what you download, and any data breaches may be attributed to you. In an astonishing case from 2021, the Queensland Uniting Church pursued charges of computer hacking against a former employee. The employee had sent emails from a work computer to the media about one of the worst paedophiles in Australian history, who had worked for them. The whistleblower was found not guilty of charges, but the distress caused was real.

A vital tactic is to create contemporaneous notes about workplace issues – regularly recording first-hand observations of problematic behaviour, power dynamics, or cultural patterns such as bullying, exclusion, retaliation, gaslighting, or ethical breaches. Make these notes on your own computer or by hand and date all entries. These notes should be factual, and preferably include direct quotes, actions taken, witness names, and your emotional response. The purpose is to create a personal log that documents systemic issues over time, helping establish patterns that may otherwise be minimised, denied, or forgotten.

Know Your Rights and Relevant Laws

It is essential to understand the legislative and regulatory environment in which you work and have a working knowledge of relevant acts of parliament [e.g. Fair Work Act (2009) in Australia, the Employment Rights Act (1996) in the UK, or the various state legislation in the US]. It will also be vital to understand, in broad terms, what whistleblower laws are enacted where you live and work. Another new set of regulations are Work Health and Safety regulations around psychosocial hazard management. In 2021, International Standard 45003 was introduced to prevent psychosocial hazards arising from the tactics we discussed in Part 2.

Be aware of the formal grievance mechanisms and any protected disclosure policies within your organisation. Remember that if they have been covering things over, these systems may have been avoided or abused. Can you find out the success rate of any previous grievances? If these grievance systems are used against you, ask for an expected timeline for the investigation to be provided in writing and make a record of each extension. When it goes on too long, a letter from your lawyer can help move things along.

Work from the Outside In

Starting your reform attempt from inside the organisation is likely to fail. There is a case to test systems for complaints, but history shows internal systems are often designed to shield the organisation rather than to find the truth. Senior leaders seldom consider that there are damaging systemic issues, as they rose to the top within this broken system. These leaders don't see anything wrong; after all, it worked just fine for them. Therefore, it is important to work from the outside in.

If gentle internal prods have achieved nothing, it's time for health and safety regulators to take action, especially if your jurisdiction has laws regarding psychosocial risk mitigation. You can often report issues anonymously. Inspectors addressing the problem will not only investigate the incident but also assess whether any systems are in place to prevent similar events. If you are part of a union or professional association, they can be a great source of support in solving workplace issues. It is best to make any report using your personal computer and to keep your anonymity.

A risky, but sometimes successful method can be to engage the media. Most of the scandals discussed in this book have been the result of a free press that supports workers' concerns. Journals as diverse as *Computer Weekly* and the *New York Post* have broken stories of corporate malfeasance that have led to *mea culpa* moments. Nothing gets an organisation to confront its issues quite like when activities they have overseen are publicly shown to conflict with its corporate values.

Finally, you can work internal channels. Not all leaders will be complicit in the perpetuation or cover-up, and some will genuinely want to see change, but their hands are tied until someone dares to pursue matters. You may also wish to see if a group of staff are willing to stand up together. One person pursuing a matter can be brushed off easily. Still, when a group of ten staff members complain together and report to diverse bodies, it is nearly impossible to avoid. If you do choose to work through the workplace, be ready for the helper to buckle under the pressure and leave you high and dry.

Also, when inducing *mea culpa* moments, beware of your own motivations. If you are driven by hate and the desire for vengeance, you are more likely to justify bending the truth. Additionally, at the end of the process, when you don't get full justice, feelings of hatred and vengeance linger indefinitely. Don't desire to get even but, instead, act out of a sense of justice and a desire for something better. Peter Fox described the right mindset in the mother of an abuse survivor, 'It wasn't unbridled anger, more an outrage restrained by reason and a resolve to see justice' (Fox, 2019, p. 3).

Preventing a Mea Culpa: A Guide for Leaders

If you are a leader or an organisation you have power to reduce the pressure and stop your workers taking the advice above. Ponder the following questions briefly. With the benefit of hindsight:

- Could Julius Caesar have done anything to demotivate his murderers?
- Could the Post Office have done anything to prevent the *Horizon* scandal and betrayal of 900 sub-postmasters?
- Could the Australian Defence Force have made a Royal Commission into the suicides of its current and former employees died by suicide unnecessary?
- Could Boeing have done anything to prevent the 737 MAX and the betrayal of its customers and employees?
- Could volunteer fire services have prevented so many *mea culpa* moments?

- Can your organisation do anything to prevent a *mea culpa* moment when its shoddy treatment of workers is revealed to the world?

The answer is 'Yes' to all these questions. However, hindsight is always clearer – the trick is to turn it into foresight. A more critical question is whether you and the other leaders of your organisation are willing to address cultures and practices that lurk in the corridors and cause moral injuries. If leaders have progressed to a stage of contemplation, this section outlines the necessary cultural changes that follow. In Chapter 10, we will bring these mind shifts into full expression through actionable frameworks that prevent betrayal, moral injury, and *mea culpas.*

Care

Avoiding a *mea culpa* moment begins with a cultural mindset shift towards genuinely caring about staff more than the organisation's halo. This care cannot be performative or paltered rhetoric, but a deeply held value that is consistently expressed through leadership decisions, priorities, and relationships. Caring for staff encompasses four interconnected elements: compassionate leadership, empathy, benevolence, and civility. Each strengthens the moral core of leadership and helps insulate against moral injury and betrayal, preventing the need for any *mea culpa* moments. Because care is central, we will dwell here longer.

Compassionate Leadership: The Compass of Care

Mark Crosweller is a former Fire Chief who draws on years of experience in emergency management. Crosweller argues that compassion is not just a virtue but a compass that can guide leaders through uncertainty and moral complexity (Crosweller, 2022a, 2022b, 2024). Compassionate leadership entails the ability to enter into the suffering of others with the intention of alleviating it. It does not mean solving each problem or absorbing every burden but instead approaching each challenge with an awareness of its human impact.

For Crosweller, compassion is not sentimentality but strength. In workplaces, this means pausing before implementing policies that may appear efficient on paper but cause emotional or moral harm. It involves being aware of how institutional pressures can silence or hurt people, and taking proactive steps to ensure leaders are standing up for those who serve in that system. Compassionate leadership signals to employees that their humanity is not an afterthought. Compassion is a compass, not a detour.

Empathy: Lessons from Hostage Negotiation

Empathy is the ability to understand and share the feelings of another – not just cognitively, but emotionally. In *Never Split the Difference*, former FBI hostage negotiator Chris Voss explains that empathy is the most powerful tool in high-stakes conversations (Voss & Raz, 2016). If empathy can help defuse armed standoffs, surely it can be applied in boardrooms, break rooms, and staff meetings as well.

Voss notes that tactical empathy does not mean agreement or surrender. It means deeply listening, naming the emotions of the other person, and signalling that you understand their perspective. It means maintaining a presence with staff, not persuasion. Applied to leadership, this means engaging employees not as problems, but as individuals to be understood. Empathy shifts culture from contractual thinking (mere legal obligations) to covenantal thinking, where leaders understand the moral 'why' of staff motivations. When leaders are empathetic, staff members are more likely to disclose concerns early, preventing minor issues from escalating into crises. Empathy, therefore, is not just a relational nicety; it prevents *mea culpas*.

Benevolence: The Heart of Trustworthiness

As discussed in Chapter 4, research identifies three core traits of trustworthy leaders: ability, integrity, and benevolence. Of these, benevolence is the standout activity that motivates trust, creating the perception that a leader genuinely cares about their people's interests as much as, perhaps even more than, their own. Without benevolence, technical ability and ethical conduct become almost meaningless traits. That's right; if you are not benevolent, staff may have little interest in how well you perform your job.

Benevolence in leadership manifests in many ways: a willingness to help without personal gain, an interest in employees' personal and professional development, and a refusal to exploit their vulnerabilities. Mayer and Gavin (2005) found that benevolent leaders are perceived as those who would never knowingly harm their employees, who care about what matters to them, and who would pause their own work to offer support. Benevolence is not about grand gestures, but instead arrives in micro-moments when managers listen intently, remember what matters to someone, or advocate for their dignity. In toxic cultures, betrayal flourishes when benevolence is absent. Trust begins to return when it reappears.

Civility: The Culture of Trustworthy Disagreement

Trust is the relational aspect of truth-telling, and one of the clearest signals of its presence or absence is how people treat one another during disagreements. Civility is not about being agreeable or polite to those like us – that is flattery. True civility emerges when people can vigorously disagree without descending into character assassination. It is the capacity to 'play the ball, not the person'. In an age where polarisation is normalised, civility is a countercultural act.

Civility demands active listening, the humility to articulate someone else's view in a better way, and a commitment to resolution rather than retribution. When this mind shift is adopted, people feel heard, valued, and respected. They then take risks, offer ideas, and voice concerns. Crucially, this engagement includes the concerns that, if ignored, become the seeds of future scandals.

These four expressions of care – compassion, empathy, benevolence, and civility – remove many of the attitudes that enable betrayal to thrive. Leaders who practice them are less likely to be surprised by a devastating *mea culpa*, because their culture encourages concerns to surface early and humanely. In place of cultures where betrayal becomes inevitable, caring leadership make trustworthiness inevitable.

Communicate in Four Dimensions

Many people who rise to leadership display the belief that their primary form of communication to workers is to tell them what to do. In my doctoral research, I found that this one-sided communication, especially when it is paltering, ambiguous, or rigid, was one of the most cited drivers of moral injury. Flaws in the command-and-control mindset can disengage staff at best, lead to poor operational outcomes in the middle ground, and, at worst, create conditions for moral injury.

In response to these concerns, I developed a 'Four Way Communication' model in which leaders adopt a new mindset about communication: listening, asking, reporting back, and only then, speaking. They form a sequence like a well-crafted 4-course meal that replaces the crassness of an all-you-can-eat buffet, at which leaders gorge themselves on simply issuing orders.

Listening: The Wisdom of Restraint

The first responsibility of communicative leadership is to listen – fully, empathetically, and often. Too many people speak first and listen last, usually just to provide an answer. But ancient Jewish wisdom tradition warns against speaking too quickly. The collection of proverbs in the Torah (Old Testament) repeatedly admonishes leaders to incline their ears to counsel. Later Christian saints reflected on this principle, saying: 'Let every person be quick to hear, slow to speak, slow to anger' (James 1:19).

Leaders who listen well create a climate where people don't just feel heard, they are heard. They become safe harbours for early concerns and moral intuitions that might otherwise go unspoken. Valid listening is active and involves more than giving a staff member attention – there is intention to understand and act on what has been heard. This intention to act is sometimes spoken of as the difference between hearing and listening. I almost always *hear* what my wife and children say to me. The soundwaves make it to my eardrums as often as those from the television show I might be watching. However, I *listen* when I don't go 'huh?' and instead act on what was said. Listening signals that what is said is valued, even when no immediate solution is available.

Asking: The Discipline of Curiosity

Listening is most effectively deployed when leaders ask penetrating questions that encourage deeper consideration of any issue and its solution. This kind of questioning is sometimes associated with the Socratic method. The questions cannot be rhetorical ones designed to reinforce authority, but genuine, open-ended questions that invite perspective, insight, and challenge. Asking should arise from curiosity rather than seeking to find blame.

Great leaders ask questions to find answers, but also to increase self-awareness in those being asked. Jesus Christ still has the greatest following of people of any historical character. Followers ascribe to him omniscience, that is, knowing everything that can be known. Yet, Jesus asked more than 300 questions of his followers – more than Socrates, who is regularly portrayed as the great question asker in history (Copenhaver, 2014). He used questions not to gather data, but to open the minds of those whom he asked. But not all questions are equal.

Chris Voss explains what active listening sounds like in hostage negotiations by introducing the concept of calibrated questions. Carefully constructed inquiries that begin with 'how' or 'what', designed to lower defensiveness, reveal motivations, and engage people in collaborative problem-solving. Questions like 'How can we make this work with the resources we have?' or 'What's the biggest challenge you're facing right now?' move beyond simplistic answers and encourage thoughtful reflection. If calibrated questions work with hostage takers, they may work with your staff.

Calibrated questions are powerful because they demonstrate curiosity without accusation, giving employees a sense of agency while allowing leaders to uncover critical information. Used wisely, they foster psychological safety and signal that the leader values what people know and have experienced. When leaders ask with integrity and calibration, they invite honesty and a shared decision-making process. Without questions, staff understand that their input is irrelevant and that the future is already decided.

Reporting Back: The Forgotten Step

Many workplace cultures emphasise listening and asking but stop short of reporting back. This omission is often where trust quietly dies. When staff raise a concern or make a request, and nothing is heard, the natural assumption is that nothing was done. The experience of being ignored is demoralising, even if a leader has investigated the feasibility of a suggestion.

Forgetting to report back to staff is the real root of 'survey fatigue'. Staff are not tired of being asked for feedback but of screaming into the void. Reporting back, even if no change is possible, is an act of transparency that allows for further listening and questioning. Leaders who report back build trust even when they cannot offer the outcome hoped for.

Speaking: The Courage to Tell the Truth

Except for immediate or life-threatening incidents, where orders need to be clear and consistent, speaking is the final part of communication. Only after listening, asking, and reporting back does a leader earn the right to speak; then, the imperative is clear: be truthful and don't palter. Don't speak in legalese, corporate spin, or as if you are hiding something – tell the truth. This is where leadership meets moral courage; truth-telling is radical and risky.

Leaders must resist the temptation to protect organisational reputation through vagueness or euphemism. First responders, for example, have finely tuned 'bullshit detectors' (Gilmartin, 2002). If trust is to be earned, speech must be honest, even when the truth is painful or incomplete. Where a Just Culture prevails (see Chapter 10), truth-telling is both safe and expected. The paradox of truth-telling is that it is simultaneously risky and safe. It risks negative consequences for past failure but opens the door to safer actions in the future.

These four modalities: listening, asking, reporting back, and speaking form the framework of morally safe communication. Each builds upon the last with well-crafted, effective words being owned and complied with by staff. When enacted together, a rhythm of dialogue is created that dignifies employees, elevates concerns, and prevents betrayal. Without them, even the most well-intentioned leader may find themselves offering a painful and avoidable *mea culpa*.

Elevate Learning to a Moral Non-Negotiable

The prevention of betrayal requires leadership with a strong commitment to continuous learning, as a moral imperative. Organisations that avoid moral injuries and any subsequent *mea culpa* moments view learning not as remediation, but as the redemption of those involved. In these cultures, learning is prized over blame, and reflection over retribution. A mindset of learning recognises the value of learning from past mistakes to prevent future failures. The other option is to scapegoat bad apples and press on as if there's nothing to see… until the *mea culpa*.

Learning involves asking not just what went wrong but how the prevailing system allowed the issue to arise. Once this learning is complete, it is possible to address issues systematically and systemically. When organisational leaders default to blame, they overlook oversights in standard operating procedures, prioritising expediency and halo protection.

In some emergency service organisations, internal investigative systems have become a source of deep moral injury. Participants in my doctoral research described internal complaint processes as unjust, punitive, and opaque. These findings echoed findings from independent reviews of two Australian ambulance services. One noted: 'The Professional Standards and Conduct Unit… is there purely to make decisions that show the least impact on the Ambulance Service, not to discover the truth and resolve the issues' (Parker, 2008). Another reported that complainants were often left without support and experienced the victimisation of retaliation (Victorian Equal Opportunity & Human Rights Commission, 2022).

These systems could be described as enabling administrative violence where policy and procedure are weaponised to shield hierarchy rather than learn lessons (see Chapter 5). This damage could be alleviated by moving from adversarial models of justice to inquisitorial ones that prioritise understanding over accusation. Law Professor Catherine Dunham (Dunham, 2022) and others argue a Just Culture does not erase accountability but refines it, aiming for fairness, transparency, institutional humility, and increased learning.

Suppose leaders wish to avoid a *mea culpa*. In that case, they must embrace learning as a moral imperative, embedding learning in the very processes by which errors, complaints, and concerns are considered. This means modelling a posture that adopts the aphorism: 'We do not fear failure, but we do fear repeating it'.

To Build Trust, Meet Less and Connect More

One of the paradoxes of modern organisational life is that while meetings proliferate, meaningful connection diminishes. Campbell et al. (2025) argue that excessive meetings can be a drain on morale by eroding perceptions of organisational support rather than enriching it. The message is clear: leaders build trust not by increasing the volume of scheduled meetings but by making time for genuine relational contact.

This view is reinforced by a report in the Harvard Business Review that documents the productivity costs of unnecessary meetings (Laker et al., 2022). Laker argues that cutting back on meetings compels leaders to communicate with greater intention and clarity while creating space for autonomy, reflection, and deeper thought. Reducing formal meetings doesn't mean reducing interaction; instead, it encourages leaders to make space for informal times of listening, co-creation, and empowerment. An incentive is that Laker reports staff productivity increased by 71% when meetings were reduced by 40% – a win-win!

However, cutting meetings is not enough. Singer and Tucker (2014) caution that well-meaning leadership practices, such as programmatic 'walk arounds' can backfire if they are perceived as performative. Safety rounds and Gemba walks, when done with sincerity and follow-through, can build trust. However, when they are superficial or lack follow-through, cynicism deepens. To build trust, leaders must shift from merely being seen to seeing their staff, and towards investing in a real-time, relational connection with those they lead. The most effective leaders do not schedule more time to talk; instead, they create space to connect.

Help Them Lead

Middle managers are frequently overlooked in conversations about leadership, often bearing the blame for failure and facing the dissonance of being caught between the decisions of senior executives and the needs of frontline staff. They are asked to implement directives with limited resources and support. When things go wrong, it is tempting to label them as obstacles to change or scapegoat them for organisational failure. However, middle leaders are not the problem; they are the solution to betrayal and moral injury.

Middle managers are the emotional barometers of an organisation. They are close enough to hear distress and pick up the early signals of betrayal, but lack the authority to enact systemic change. This tension can leave them demoralised and burned out. While middle managers may receive technical training, few receive formal training to navigate the moral and relational complexities of their role. Money could be spent on leadership development rather than individual wellbeing programmes as a better way to reduce rates of moral injury and other comorbid mental health conditions.

Research is clear: leadership style matters not only for performance, but also for psychological safety and wellbeing. Paramedic and researcher Brian Bayani found strong links between poor leadership behaviour and moral injury in first responders (Bayani, 2021). Further, Danielle Mercer and her team reported the importance of relational and context-sensitive leadership among paramedics at the frontline (Mercer et al., 2018). These findings underscore that the middle-level leaders relied upon to maintain stability in times of crisis are often those least equipped to do so.

To avoid a future *mea culpa*, organisations must start treating leadership investment as a cultural necessity rather than an executive privilege. Professor Jacqueline Drew argues that 'leadership for wellness', where moral injury and other mental health issues are prevented, is possible but requires a system-level commitment (Drew et al., 2023). Helping middle leaders lead requires deliberate training, supervision, and providing them time to reflect. It means recognising that the road to moral injury prevention through moral resilience runs primarily through those in middle management. When we help them lead, we create systems that not only do not harm but also heal.

The five mind shifts – caring, communicating, learning, building trust and connection, and enabling leaders – provide a foundational shift, but leaders must choose whether to act. When these culture shifts are neglected or treated as optional, the costs are rarely immediate; instead, they are cumulative. Betrayal takes time, and moral injury accumulates one small cut at a time. But when the reckoning comes, it is loud, wide-ranging and expensive.

Choose Your Costs

Organisational leaders know that nothing comes for free, but costs differ from one another. Wise and diligent leaders don't just seek to minimise expenses, but also balance them against risk, consider opportunity costs, and assess return on investment, among other factors. There will be a cost if you choose to overtly prevent morally injurious cultures through implementing the recommendations above. There is another cost for delayed action. Below is a summary of the costs for organisations that did not act early.

The failure to address endemic illegal behaviour at Enron resulted in USD \$74 billion in investor losses and USD \$7.2 billion in settlement as well as imprisonment for senior leaders.

The three scandals that engulfed Qantas saw a drop of around 28% in annual profit (AUD \$500 million), AUD \$300 million in compensation and fines, as well as the premature exit of its CEO, which included non-payment of his bonuses.

The Boeing 737 MAX fiasco cost approximately USD \$20 billion due to the worldwide grounding of aircraft, plus USD \$1.1 billion in settlements. These figures do not include lost orders and decreased market share to its main rival, Airbus, with whom the cost-cutting measures were designed to make it more competitive.

The Robodebt scheme cost the Australian Government approximately AUD \$1.2 billion in payouts and AUD \$112 million in a class action. The Post Office *Horizon* scandal has cost UK £1 billion in compensation and over £50 million in inquiry costs so far.

The enormous financial costs of failing to prevent these kinds of incidents are clear – costs to avoid betrayal are large, but insignificant in comparison. However, the human cost is greater than the dollar amounts. It is common knowledge that psychological injury claims are increasing around the world, attributed mainly to the morally injurious results of organisational betrayal. The case to act is clear, but are leaders ready to act?

POINTS FOR CONSIDERATION

- What are the simmering tensions in your workplace that may be seeds of a *mea culpa* moment? Are you the right person to induce or prevent a *mea culpa*?
- If you are seeking a *mea culpa*, take time to reflect on your motivations. Are you acting out of hatred and revenge or a desire for justice?

9 Saving Ourselves
Defending Virtue

A true word.

> *If you can wait and not be tired by waiting,Or, being lied about, don't deal in lies,Or, being hated, don't give way to hating,And yet don't look too good, nor talk too wise; [...]If neither foes nor loving friends can hurt you;If all men count with you, but none too much;If you can fill the unforgiving minute with sixty seconds' worth of distance run—Yours is the Earth and everything that's in it...*
>
> Rudyard Kipling

INTRODUCTION: IN A NEW YORK MINUTE

December 4, 2024, was a cool, calm, and overcast morning in New York City. About 6:45 am, two strangers from diverse upbringings encountered each other on West 54th St. One was a child of wealth, a private school valedictorian and Ivy League graduate; the other was from a humble, even impoverished, rural background. Nonetheless, this second man had been valedictorian at his rural public high school. One was a CEO of a large US corporation, the other a seemingly restless wanderer who travelled cheaply by bus to New York and lodged in shared accommodation in the days prior to their encounter.

After the early morning encounter, the CEO lay dead from gunshot wounds. At first glance, this is yet another tragic shooting following a predictable script of urban violence in the US. However, the one who pulled the trigger was the privileged Ivy Leaguer, while the CEO was a rags-to-riches tale that led him from a humble rural town in Iowa to the heights of corporate America.

Brian Thompson was CEO of United Healthcare, America's largest health insurance provider, with an annual profit of around USD $22 billion at the time of his death. Corporate America, particularly the insurance sector, have been heavily criticised for a seemingly heartless drive towards enormous profits at the expense of clients' welfare. Both traditional media and popular culture have particularly criticised its insurance practice of denying or delaying legitimate claims.

As mentioned in Chapter 5, John Grisham's novel, *The Rainmaker,* gives artistic smoke to the fire of heartless insurance companies and is informative in the tragic

 DOI: 10.1201/9781003499725-12

New York encounter. The alleged murderer, Luigi Mangione, had suffered for years with a painful back condition that some have speculated may have raised a sense of grievance at the insurance sector. Rumours also grew because bullet casings found at the scene reportedly had the words 'delay' and 'deny' written on them. These words have come to symbolise the tactics of Grisham's *Great Benefit Insurance* and summarise the actual practices of US health insurance companies. Notations in Mangione's diaries detail his grievance at the profit-making of the health insurance industry. Mangione perhaps felt betrayed and abandoned by the insurance industry to whose apex Thompson had risen.

One might imagine the backgrounds of the men to be reversed. This strange tale prompts us to consider what lies behind the divergent paths these men took. It is particularly important to ask what led someone with every advantage to kill the very person he could have become. The encounter tells a tragic tale of how perceived betrayal may lead to the degradation of moral virtue and, more alarmingly, a society's moral vision.

Whatever is finally revealed about Mangione's motives, what is certain is that many people around the world saw this murderous encounter as a moment for celebration or at least with some schadenfreude. Social media, long the open sewer of human communication, was alight with delight and justifications for this murder. Stories poured out about the corporate cruelty of insurance companies and how Thompson 'had it coming'. Adding to this online melee was the delight or equivocation of seemingly respectable commentators.

Bret Stephens of the *New York Times* wrote that one ex-Times reporter said they felt 'joy' at the murder of Thompson, while a US senator said that while 'violence is never the answer', the murder arose because 'people can only be pushed so far'. This case, like Brutus' murder of Caesar, is an example of the terrible actions people justify and encourage in response to perceived organisational betrayal.

Considering the celebratory feeling around the murder, it is worth asking if this response to betrayal of moral principles is what we want for our society. Is the course Mangione took one that any of us might wish to take? I for one answer, 'God, I hope not!' The taking of Thompson's life and the ensuing celebration of his murder reveal a society-wide case study of what Ned Dobos (2015) calls character degradation.

OUR MORAL FIBRE

Moral philosophy has a history as long and storied as that of the examination of truth claims (epistemology), detailed in Chapter 6. We will not spend much time rehearsing the history of moral philosophy, but will instead note that 'virtue ethics' will be our guiding light and foundational assumption in this chapter.

Virtue ethics holds that our ethical (or unethical) actions are neither a function of external compulsion or authority (deontological ethics), nor arise from a belief that actions should solely be judged on their outcomes (utilitarian ethics). Instead, virtue ethics holds that our moral reasoning stems from internal virtues, sometimes summarised as 'character'. In reality, most of us choose our moral actions based on a mix of deontological, utilitarian, and virtue ethics.

The highlighting of virtue ethics, however, helps us understand how a moral injury might affect our ability to discern right from wrong and then enact it. The moral impact on those betrayed has been an often-overlooked strain of moral injury research. The seminal work on moral injury by Jonathan Shay was titled 'Achilles in Vietnam'. What many people omit, to our loss, is the subtitle: 'Combat trauma and the undoing of character' (Shay, 1994). Unlike post-traumatic stress disorder (PTSD), moral injury often entails a destruction of the very same moral code that has been injured. In short, moral wrongs done to a person can serve as a personal justification for committing wrongs to others.

Shay elaborates on this theme, particularly highlighting how betrayal by leadership during war can undo character, not just psychological damage, but moral disintegration. Shay writes, 'The soldier's character – habits of mind and heart essential to functioning in society – can be undone' as a response to trauma and betrayal (Shay, 1994, p. xx). The erosion of social trust, cooperation, and citizenship displayed in response to Thompson's murder is a potent civilian symbol of society-wide moral degradation.

It must be stated firmly that the undoing of personal character that often attends moral injuries requires our personal attention if we want our organisations to improve. Chapter 1 channelled Tolstoy's belief that the key to changing the world is a 'permanent revolution — a moral one; the regeneration of the inner person'. The good news is that the suffering associated with betrayal can renew us, rather than destroy. However, before exploring that note of hope, let's unpack the slow process of moral degradation.

Descending into the Abyss: Moral Degradation

Moral degradation does not typically occur in a single moment, but rather through a series of disillusionments, compromises, and moral concessions. This process can unfold over months or years, particularly when one's moral purpose is strong, but the prevailing culture and practices betray that purpose. Degradation is also slow because it seeps in through ignored warning signs and tolerated systemic dysfunction, which combine with our seldom-examined personal moral covenants. When individuals attempt to maintain virtue and are met with betrayal, paltering, or scapegoating, they often undergo a progressive undoing of character. Five stages of moral degradation can be observed: naivety, caution, cynicism, resentment, and entitlement. Each of these steps erodes moral integrity, sowing seeds for the next stage with ever-accelerating impact on one's virtuous moral character.

Naivety: The Glow of Purpose

When individuals first enter mission-driven movements or organisations, they often carry a sense of moral clarity and optimism. Young teachers long to see children flourish, bright-eyed lawyers love the thrill of justice, and graduate engineers are ready to build safe and productive community assets. This 'naivety' is not foolishness, but a necessary openness to purpose, values, and the belief one can make a difference. Failure to have a period of innocence is itself a matter for concern. Our naivety is deeply rooted in the development of our moral covenants and vocational ideals – the moral horizon of the self, as Charles Taylor (1992) would put it. Many of us may refer to it as a sense of calling or passion.

However, much like love opens us to betrayal, optimism makes us vulnerable to the loss of innocence. As Shay observed, the betrayal of what one believes is right by those who were meant to protect our moral framework can shatter one's sense of self. Naivety can lead to profound disillusionment if moral expectations are unrealistically high. I spoke to a paramedic who vowed early on, 'no-one will die in my ambulance'. This expectation was quickly broken because it was utterly unrealistic. He was spared a moral injury because he was able to reflect and revise this expectation.

Our discussion on moral covenants (Chapter 2) is worth reconsidering here. Was the agreement you tacitly made with yourself and your organisation realistic? Many years of supporting emergency workers have shown the greatest distress arises not from lies and broken legal contracts, but from a combination of corporate paltering and broken moral covenants. When the organisation's practices fall short, the resulting injury feels like a betrayal not only of principle, but of identity. So, reflect on what things really made you believe the initial naïve view of your workplace. It could be that a re-assessment of our expectations can restore a sense of moral coherence.

Caution: The Subtle Withdrawal

Should you look back and assess that your expectations and moral covenant were reasonable, and things remain grating in your organisation, you may move on to the second stage. A small and cautious withdrawal from the eagerness of our rookie days marks the second stage. Individuals notice that the organisation is not what they initially hoped. Talk of openness and collaboration is a veneer for internal corporate politics, avoidance of hard truths, and ethical compromises. Here, people become cautious: they reconsider raising concerns and start second-guessing whether they should speak out. They may also reduce any extra physical effort or moral diligence at work in case another moral curveball hits them.

Much like naivety, caution is not necessarily a bad thing and can be a helpful survival strategy. When employees sense that a workplace is less safe than expected for open disclosure, or that leaders are too busy measuring KPIs to provide guidance and follow-up, they shift from contributing to self-protection. They still care, but start to compartmentalise tasks, withdraw emotionally, and reconsider if workplace systems and colleagues want or value their input or their presence.

What are you cautious about in your workplace? Be specific and name the person or process concerning you. Don't do this in the workplace just yet, because each of us needs to understand what is going wrong fully. When you can identify what is causing you hesitation, discuss it with a trusted friend or mentor to explore potential next steps. Caution may be appropriate because we cannot reasonably expect all leaders and systems to meet every moral expectation. However, caution can be a step to something more sinister. Virtues of honesty, courage, and hope begin to fade through almost unnoticeable acts of omission, which, if not addressed, can lead to destructive emotions and actions.

Cynicism: The Unhealthy Norm

The third stage is cynicism – typical, right? Cynicism marks the first overt signs of moral degradation; extending past mere self-protection, it starts to shift one's worldview. Former Sheriff and police psychologist, Kevin Gilmartin, notes that police wellbeing issues are not about the terrible things they see, but the impact on the way they see the world (Gilmartin, 2002). In the caution phase, dysfunction is a surprise, while in cynicism, dysfunction becomes the norm. Cynicism does not respect the end of your shift; you will take this cynicism home and inflict it on your loved ones.

It is reasonably easy to recognise the arrival of cynicism through our response to something even mildly unpleasant. People say, 'Typical', or 'That's just how it is'. In Australia, it has become common to say 'it is what it is' as a form of resignation to less-than-ideal circumstances. I worked in one organisation where it felt like 'it is what it is' was the realistic organisational slogan. However, when I worked in the dilapidated headquarters of the organisation, I learnt the motto there was 'you get used to it'. During cynicism, verbal irony, sarcasm, and resigned humour become the currency of conversation, signalling that something worse is likely to happen if a glimmer of hope is not restored. Gilmartin bluntly believes cynicism is measurable through the number of times one says bull*^%t in a day.

Cynicism is not just emotional fatigue but one of the three core symptoms of burnout, alongside emotional exhaustion and perceptions of reduced self-efficacy (Maslach & Leiter, 2016). Burnout in mission-driven occupations often involves moral conflict alongside issues such as workload stress. Cynicism tends to numb moral sensitivity, allowing people to work without immediate emotional cost. However, cynicism is corrosive on our loved ones, our work performance, and our long-term wellbeing and moral character. Cynical individuals tend to disengage from moral reasoning and shift towards moral relativism.

You may not suspect you are cynical, but assert you are realistic, but others will know what you only suspect. Ask someone you really like and trust to record how often you use the linguistic markers of 'typical', 'that's just how it is', 'you get used to it', or 'bull*^%t', etc., on a weekly, daily, or hourly basis. The more regular these utterances, the more accustomed you become to them. If they are daily utterances, you will need the help of a trusted ally or professional help to prevent progression to burnout or the next stage of moral degradation.

Resentment: The Rise of 'Theyism'

When thrown into the witches' brew of poisonous workplace cultures and practices, cynicism often gives way to resentment. In this stage, the organisation and its leaders are not seen as disappointing or frustrating – they are the enemy. A sign of this progression is heard in the use of 'they' to describe management, HR, or leadership. This depersonalising language signals that a person no longer feels part of the whole but is in opposition to it. When a person continues to use 'our' in their discussion about workplace matters, they still believe they're part of the bigger entity, while 'they' shows an increased distance between them and the ever-growing group of people who are perceived to have betrayed them.

In my research, I asked 227 respondents how many times leaders, systems, and colleagues had betrayed them during their career. Sixteen per cent reported being betrayed more than 200 times, while 37% said they had been betrayed more than 50 times. I am doubtful that people actually have the time to be betrayed more than 200 times, but I am sure that people perceive this to be the case. There comes a tipping point, close to the onset of resentment, when every action taken by 'them' (management, HR, PR, legal, etc.) is automatically coded as betrayal, with *post hoc* rationalisations providing all the evidence needed to allow the self-fulfilling prophecy to become 'my truth'.

During resentment, people tend to typecast specific colleagues or supervisors as betrayers, and moral language becomes retributive: phrases like 'they don't care', 'they just don't get it', or 'they deserve what they get' become more frequent. At this point, a parade of seemingly justified workplace rebellions starts. Acts of malicious compliance, poor organisational citizenship behaviour, presenteeism, and absenteeism become acts of protest. In the late stages of resentment, what began as personal hurt seeks to inflict harm on others. Caution is the sowing of moral degradation, cynicism is budding, and resentment is growth, but entitlement is the flowering of moral degradation.

Entitlement: Becoming What You Hated

The final step in this descent is entitlement. Here, individuals feel morally justified in breaking rules, taking shortcuts, or enacting retribution. They may feel entitled due to all they have suffered and therefore take liberties they would once have condemned in others. This is the tragic irony of moral injury: it wounds but also degrades our heart and soul. Finally, things come full circle with the morally injured becoming the betrayer. In this phase, people may start misusing time, resources, or power, even justifying the use of physical violence. They may justify lying, sabotaging colleagues, or refusing to help colleagues in proportion to their own perceived suffering.

A profound tragedy is that entitled people think they are acting morally by balancing the scales of justice. But in doing so, they become the kind of person they once entered the organisation to protect others from. Worse still, this behaviour often triggers justified disciplinary processes against them, further confirming the sense of betrayal and alienation.

It is almost impossible to recover from this position unless a person has their own personal *mea culpa* moment. A partner will move out, colleagues rightly fear vengeance, and internally disciplinary action becomes inevitable. What began as a breach of our unwritten moral covenant by a leader becomes a breach of legislation by an employee. Issues that the organisation paltered become matters of obsession, delusion, and outright dishonesty by the morally injured and entitled employee.

Can the Descent be Interrupted?

Moral degradation is not inevitable, but is highly likely in institutions that fail to guard virtue. The descent from naivety to entitlement doesn't just erode individuals; it changes their relationship to the world and further diminishes organisational culture. If moral injury begins as a betrayal of what is right, then moral degradation is the slow death of one's will to do right. The following section explores what good looks like and tactics to climb out of the abyss of betrayal and its hellish consequences.

In fact, if we are deliberate, suffering and betrayal can lead to the making of character. Almost all of my better character attributes have developed through hardship, and the impact of suffering has been a catalyst for character development and hope for the better. The author's most profound hope is that readers will experience growth in moral character, rather than the decline evident in many who are morally injured.

A Better Story: Moral Growth

In growth are the elements of Tolstoy's personal moral revolution through taking our moral fate into our own hands. In the battle for our character, many of us are starting from a position of weakness. Of deep concern now in progressive cultures is a significant retreat from personal self-efficacy, moral or otherwise. While humans have always blamed others for their failings, this trait has evolved into an industry that promotes blaming external factors for one's lot. Researchers call this a switch from an internal to an external locus of control. At an ever-growing scale, people are painting themselves as the victim of circumstances over which they have no control.

I want to clearly state that we each have more agency than we realise, particularly when it comes to our moral attributes and actions. We are never forced into murdering CEOs or celebrating it. Nor are we helpless spectators to our own milder forms of vengeance-taking. Better comes through a four-part strategy for personal moral growth.

Moral Growth: Building Strength from Sorrow

While moral injury and betrayal can have a degrading impact on our character, they do not dictate our destiny. In fact, a growing body of research has demonstrated that suffering, particularly that which disrupts our moral worldview, can be a catalyst for post-traumatic growth (PTG). Growth is preferable to the notion of resilience, which is the capacity to endure hardship and remain unchanged. As psychology researchers Tedeschi and Calhoun describe, growth is a movement to a different and often deeper way of being (Tedeschi, 1996). PTG describes the transformation that can occur because of hardship. Tedeschi and Calhoun reported this occurs in five domains:

- Increased appreciation of life.
- Improved relationships with others.
- A sense of increased personal strength.
- Recognition of new possibilities.
- Spiritual or existential growth.

Moral injury is a shattering of meaning, but it also clears space for new meanings to emerge. In my own research, many of the morally injured people I interviewed did not simply suffer; they re-evaluated, re-narrated, and in some cases, recommitted to deeper moral values. Some left their organisations but maintained their integrity. Others stayed and became quiet reformers, modelling the change they wished to see. In both cases, the betrayal became a starting point for moral growth instead of a slow moral degradation.

Richard Tedeschi has argued that PTG is most likely when people are given space for reflection, dialogue, and reintegration. This means time to process the experience, moral vocabulary to articulate it, and the social or spiritual frameworks to place it in a meaningful narrative. Importantly, PTG is not guaranteed; it requires work. Growth does not happen as a direct result of trauma, but rather through the struggle to find meaning in the aftermath (Tedeschi, 2004).

Part of this struggle involves literally or metaphorically rewriting your personal narrative. I say 'literally', because part of my research involved analysing 21 autobiographies of first responders. Some were explicit that writing their story brought growth and healing, giving them power back over their own life story that betrayal and trauma had taken away. When people are given tools to re-story their experiences, they can reframe moral collapse into moral clarity. This becomes a form of identity reconstruction, which entails a shift from being a victim of betrayal to a witness of truth or a bearer of virtue.

Furthermore, Tedeschi's later work highlights the importance of spiritual transformation as a domain of PTG (Tedeschi et al., 2017). This is particularly relevant in workplaces of service or mission, such as healthcare, chaplaincy, or first response. When people are betrayed in these environments, it is not just their professional selves that are wounded, but their spiritual cores as well. Yet this core can also be where seeds of renewal are nurtured. A recommitment to values, to service, to community, and to transcendence is possible, often times stronger because of the fall. Moral growth, then, does not ignore suffering – it feeds off it to produce a different kind of flowering than one of degradation.

Moral Direction: The Maltese Moral Compass

If moral growth follows the rewriting of our story and recapturing of our moral code, then it is good to ask: what sort of people should we become? What attributes are worth cultivating in the aftermath of betrayal? What character traits do we need to prevent moral collapse and promote moral repair? One source of inspiration is the Maltese Cross, which has long been associated with the Knights Hospitaller and is emblazoned on the uniforms of many paramedics. The Maltese Cross is more than a decorative or religious symbol; it is a moral compass.

The four branches of the cross represent: faith, justice, temperance, and fortitude. These character traits are somewhat dour compared to those we see on reality television. Some may wonder if they want to possess temperance and fortitude, but we all want a friend with them. When I look at paramedics wearing this symbol where I serve as a chaplain, I see some of the most profound and caring people I know. These traits speak to the depth of their character, forged under challenging circumstances. Faith, justice, temperance, and fortitude are forged in the crucible of suffering from betrayal and moral injury. If one deliberately cultivates these traits, they grow in depth and wisdom.

Further, each of the four arms has two points, making eight points in all which historically have represented a virtue critical to moral character, particularly in service to others:

- Loyalty,
- Piety,
- Frankness,
- Bravery,
- Glory and honour,
- Contempt of death,
- Helpfulness towards the poor and sick,
- Respect for the church.

I will address the last item first, as many will rightly squirm at this idea of respecting the church. Throughout this book, I have spoken bluntly about the church's cover-up of abuse, something of great shame. It is also true that many in the Western World have moved some distance from organised religion. Regardless of any religious status, humans maintain a spiritual sense of meaning and identity, as explored in Chapter 1. For some, then, this last point could instead reference their own guiding spiritual principles through which they make sense of the universe and their place in it. For others, it may be commitment to the mission and ethos of your organisation as stated, even if it is not fully lived out. Looking beyond this point, these virtues provide solace to the pain of moral injury and a true moral north to avoid being lost in moral degradation. So, let's explore some of these more closely.

Loyalty/Faithfulness

Loyalty is a two-edged sword as it is utilised to galvanise trust and enforce compliance. This attribute describes not blind allegiance to a person, regardless of their moral degradation, but covenantal faithfulness to ideals that transcend any relationship. In the aftermath of betrayal, one's capacity for loyalty and faithfulness can be deeply shaken. When we become trustworthy and faithful individuals, we can become sentinels of recovery for others. Loyalty grounded in conscience rather than hierarchy sustains communities of integrity.

Bravery and Contempt of Death

While few modern professionals face literal death at work, the metaphorical courage required to stand alone, to challenge injustice, or to act when it is unpopular is no less demanding. Those recovering from moral injury often feel intense fear of isolation and reprisal. However, bravery is not the absence of fear; it is courage to stand up when racked by it. Recognising one's own bravery in standing up against organisational dysfunction and betrayal enables people to reclaim their moral agency in the face of systems that previously silenced them. One of the things those who experience moral growth come to understand is that they have already overcome terrible suffering, so they can do it again.

Helpfulness and Compassion

Moral injury often emerges in systems that betray the vulnerable, destroying the will of people to be compassionate to others. However, many trauma survivors report an enhanced commitment to others' welfare, especially those who suffer. Likewise with moral injury, moral sensitivity becomes what philosophers like Martha Nussbaum call 'compassionate justice', which is a form of redemptive service (Pyy, 2021). Recognise that your dysfunctional workplace may not be the place to express not less and not more than is required. For example, when medical staff feel unsupported, they may take it out on patients, something to be avoided. However, when the clinician finds other outlets to demonstrate this compassion, such as through community volunteering, they can also return to a state of appropriate compassion in the workplace.

Frankness (Explicitness)

In light of our discussion on paltering, this virtue is fundamental in the moral repair process. Betrayal often thrives in cultures of secrecy and half-truths. Frankness, or what the chivalric tradition sometimes called explicitness, means speaking plainly, naming wrongdoing, and not allowing moral rot to hide behind organisational bureaucracy or personal entitlement. It is the moral courage to be direct and honest when it counts. Frankness, when combined with civility and the desire to learn, is a powerful personal and corporate virtue.

The Maltese Cross becomes more than a historical emblem; it is a moral compass. It names the attributes that betrayal can destroy, but which we can consciously reclaim. These virtues cannot be taught in a training course, captured in corporate codes of conduct, or measured by performance metrics. They are forged in difficulty, refined by reflection, and acted on with purpose and specificity.

Moral Self-Efficacy: Becoming a 'Specific Survivor'

One of the most damaging myths in trauma and betrayal recovery, is the idea that healing is passive – that time alone will 'heal all wounds'. Kevin Gilmartin insists in his work with law enforcement and emergency workers that people who genuinely recover from adversity are not those who wait to be healed but those who take specific action. He calls them 'specific survivors' (Gilmartin, 2002).

In his influential book *Emotional Survival for Law Enforcement* and a later article, *Keeping Good People Good*, Gilmartin distinguishes between those who drift into bitterness or disengagement and those who make conscious, proactive decisions to live differently in the aftermath of betrayal (Gilmartin, 2014). The specific survivor, he says, takes ownership of their recovery and their moral life. This is a radical assertion in cultures that increasingly shun suffering and externalise accountability.

Gilmartin identifies several hallmarks of specific survivors. Specific survivors anticipate the emotional impact of betrayal and plan for it. Their initial naivety must be balanced with a sense of realism. Rather than being blindsided by institutional dysfunction or colleague disloyalty, they come to understand it as a risk of service-based work. This mental framing does not eliminate injury, but allows for swifter moral recovery.

Specific survivors build and maintain emotional energy outside of work, developing friendships, hobbies, and habits that ground them in meaning-making activities not derived from their professional identity. Having a broad basis for meaning-making is vital in an era when people often derive their personal meaning and identity largely from their job.

Specific survivors practice principled defiance or 'outrage restrained by reason and a resolve to see justice' (see Chapter 8). That is, specific survivors know when to push back against policies or cultures that violate moral values, doing so with clarity and composure rather than rage or retribution.

Specific survivors resist the victim narrative. Gilmartin is blunt about this: trauma and betrayal happen, but wallowing in helplessness or outrage is not a recovery plan. Instead, specific survivors act in ways that align with who they wish to be, rather than the system that has wounded them. The challenge here is to stop waiting for institutional change or the betrayer to apologise. The specific survivor focuses on the domain they can control: their own moral conduct.

In my own research, several participants who faced betrayal reported a moment when they snapped out of passive suffering and made conscious decisions to reclaim their agency through career changes, speaking out, deepening spiritual practice, or mentoring others. Such action is the mark of a specific survivor: not a return to normal, but moral flourishing. Moral recovery is not an aspirational wish but a set of choices that the specific survivor makes.

Moral Practices: Depth and Breadth

If betrayal and moral injury undo character, then the road to moral renewal must be walked with specific intent and time for reflection. Selecting virtues to pursue and acting with intent are vital. There is also a need to stop and reflect at regular intervals along the way through explicitly moral reflective practices. Moral reflection involves a set of habits and disciplines that build moral character over time and help us detect early signs of straying from the path. In this final section, we consider two key moral practices that sustain and deepen character in the aftermath of injury: moral imagination, which provides a depth of understanding, and enlarged thinking, which broadens our moral horizon.

Moral Imagination

In a seminal article on moral judgement, Mark Johnson describes moral imagination as the capacity to envision the whole of humanity and the moral stakes of any situation (Johnson, 1985). Moral imagination sparks the possibility of continual growth in character in a way that simple adherence to rules and phony corporate ethics statements seems to extinguish. Moral imagination involves taking stock of who we truly are by delving deeply into the issues we face in the workplace and throughout our lives.

There is a recursive relationship between moral imagination and morally upright action. Ethicist and theologian Andrew Cameron holds that many people leave reflection until moments of crisis and find themselves wanting due to a lack of moral self-awareness. Cameron says our moral code is like an iceberg; we see only a small part of it in our day-to-day activities, but when a crisis occurs, we need the great underwater depth of our moral iceberg. Because of a shallow understanding, we might stray into poor moral choices (Cameron, 2011). On the other hand, what is most eroded in morally injurious systems is not merely a sense of right and wrong, but also our ability to flex our moral imagination. People begin to believe there are no alternatives – the system is fixed, everyone is corrupt, and 'this is just how things are'.

In the face of morally complex situations faced at work, some form of structured reflection enhances moral clarity and builds character. An older form of structured reflection is the five-step reflective practice called the Ignatian Examen, which helps individuals pause daily to explicitly notice, evaluate, and respond to events.

The Examen is grounded in Jesuit spirituality but has been widely adapted across secular and various faith traditions due to its usability (Fuchs et al., 2023). The steps are: presence, gratitude, review, reflection, and response. These enhance moral discernment by helping individuals reflect on how their thoughts, feelings, and choices affect others and align with their deeper values. The Examen can help practitioners in morally injurious contexts to reclaim agency, recover a sense of coherence, and strengthen the moral fibre often frayed by betrayal or burnout.

Chaplain and researcher Rev Dr Sally Nash builds on the examen through the practice of *regenerative reflection*, a model designed to help individuals reconnect with their core ethical commitments after experiencing moral harm or ethical dissonance (Nash, 2011). The practice can help us answer 'what happened?', but also, 'Who am I becoming?', and 'What do I stand for now?' Moral reflection of this kind is both backward-looking and future-oriented, making it an integrative practice that helps us forge coherence and purpose after disruption.

A trio of policing leadership and moral injury experts has produced *The POWER Manual,* which includes a convenient version of moral imagination (Blumberg et al., 2022). In their guide to spiritual and moral fitness, they first ask police to create their own mission statement, containing aspirational principles and outlining what they hope to achieve through their actions. The second and third steps involve listing on small cards their personal values and the places where meaning and purpose are found. The fourth step is to foster contentment. Importantly, they recommend setting up regular times to review not just our failures, but also when things have gone well. A good question that can be asked for positive reinforcement of our moral action is, 'What would have happened to that person if I hadn't turned up today?' Such questions allow congratulation and not just moral self-censure.

Moral imagination resists despair and asks: what would it look like to act differently? What might mercy, courage, or honesty require in this moment? It rehumanises both the victim and betrayers without excusing wrongdoing. In my research, many redemptive moments occurred when participants stepped out of expected responses and made deeply human, creative moral decisions, such as small, quiet acts of care or truth-telling that resisted the culture that they may feel oppressed by. These were acts of imagination as much as conscience.

Enlarged Thinking

The second key moral practice is what theologian Miroslav Volf calls enlarged thinking (Volf, 2019). This refers to the discipline of thinking with a moral breadth that extends beyond the horizon of one's moral code, one's own tribe, and one's own grievances to intentionally understand the moral perspectives of others, especially when they are one's enemies.

Volf, drawing on both theological and philosophical traditions, argues that moral maturity requires more than clarity; it requires breadth. In his book *Exclusion and Embrace*, he contends we must be able to make space within ourselves for the voices of others. This breadth is central to refusing to 'other' or even 'they' those who betrayed us. Enlarged thinking does not mean absolving harm but rather resisting the moral shrinkage that occurs when we reduce our betrayers to caricatures.

Practising enlarged thinking involves: listening to dissenting views without immediately defending one's own, recognising situational pressures that shape the moral failure of our enemies. This process still leaves space for appropriate recompense to be paid by violators, and space to reflect on how one's own pain might narrow or distort our understanding of those who betray us.

Enlarged thinking is an individual worker's complement to a leader's empathetic listening to them, which was central to the organisational mindset shift (see Chapter 8). There, empathetic interactions with workers were key to sound leadership, and here, enlarged thinking allows the leader's lot to be thoughtfully considered. In this recursive system, should it ever be fully achieved, empathy breeds workers who exercise enlarged thinking, and enlarged thinking encourages empathetic leadership.

When this system works, it fosters interpersonal and corporate reconciliation, replacing the broken social trust that is the hallmark of betrayal and moral injury. Enlarged thinking is the practice that stops survivors from becoming perpetually aggrieved or morally isolated. It invites them into a larger moral story, one that encompasses both justice and grace, as well as truth and forbearance.

As you write your narrative of workplace betrayal, it can be a destructive tale of vengeance and hatred, or it can be a transformative and redemptive one. The positive trajectory that moral imagination and enlarged thinking enable does not diminish the significance of paltering and scapegoating; rather, it brings a crescendo of connection and a resolution of personal peace. By reclaiming the virtues of the Maltese Cross, becoming specific survivors, and deliberately practising moral imagination and enlarged thinking, we begin to write a better story, not just of our own recovery, but of the kind of society we want to be part of. And perhaps, in doing so, we might reclaim not only our integrity, but our hope.

The choice is yours as to what story you want to write. Write it well.

ITEMS FOR CONSIDERATION

- What verbal or active signs can you see in yourself that you may be in some stage of character degradation? Are your words cynical? Are your actions vengeful?
- What stage of degradation might you be in, and what steps can you take to reverse this?

- Which of the elements of moral growth is most appealing to you? Choose one, and then develop time and strategies to implement it.
- If you have betrayed the trust of someone in the past, what would you want them to know? For those who have betrayed you, what might they want you to know.

10 Saving Our Organisations
Courage and Curiosity

A true word.

> *The road to hell is paved with good intentions.*
>
> Unknown

> *In the end, reconciliation is a spiritual process, which requires more than just a legal framework. It has to happen in the hearts and minds of people.*
>
> Nelson Mandela

INTRODUCTION: ENACTING CORPORATE MIND SHIFTS

In the house where I grew up, a circular piece of paper with the word 'tuit' in large letters was held onto the refrigerator with a magnet. Tuit is not a real word, but you possibly get the joke. When visitors inquired about the paper, the answer would come, 'It is a round tuit, have a look'. On grabbing the paper, they would eventually turn it over and see this text,

> This is a round tuit. Guard it with your life!
> Tuits are hard to come by, especially the round ones.
> It will help you become a much more efficient worker.
> For years you've heard people say "I'll do that when I get a round tuit."
> So now that you have one!
> You can accomplish all those things you put aside until you got a Round Tuit.

Of course, this was a subtle jibe from my parents to the lazy teenagers my siblings and I always flirted with being. However, it is relevant to this chapter. Having detailed the nature and mechanisms of workplace betrayal, Chapter 8 listed a series of corporate mind shifts necessary to prevent a corporate *mea culpa*. However, corporate leaders might be tempted to treat those mind shifts like corporate ethics statements – lovely ideas that are never put into action. In Chapter 9, individual mind shifts and actions to safeguard oneself were presented in a single package. Now it is action time for organisational leaders. In the spirit of spurring good intention to action, this final chapter becomes the organisational 'round tuit' of the book.

DOI: 10.1201/9781003499725-13

The proactive attention by senior leaders to the matters raised in this book is vital because betrayal can largely be prevented. Through the elimination of destructive practices and cultures and filling the space with a systemic approach to prevention, harm can be reduced. The prevention of moral injury cannot, and indeed should not, be the sole responsibility of individuals; it requires a systemic approach. Organisational leaders have a mandate and a moral imperative to act on moral injury. Four systems frameworks are now recommended, each with practical and actionable recommendations that can make your workplace pleasant as well as safe and productive.

BUILDING INSTITUTIONAL COURAGE: REVERSING THE HARMS OF ORGANISATIONAL BETRAYAL

The Royal Commission into Defence and Veteran suicide in the ADF had 122 recommendations. Ten per cent of the recommendations were about crimes of sexual assault against ADF personnel by ADF personnel. What happened to far too many survivors was a profound betrayal of trust by military leaders and the cultures those leaders created, curated, and thrived in. The relevant recommendations sought action for the sexual assaults, the cover-up, and the scapegoating of survivors.

It is in this context of sexual misconduct that the notion of organisational betrayal and its antidote, institutional courage, was born. After reporting a sexual assault in her college dorm, a 19-year-old freshman was threatened by other students, while the pleas of her parents to the college were ignored. The young lady tragically died by suicide 10 days after reporting the crime. The student against whom the allegation was made did not miss one game of football for his college. Her father later called it a betrayal by a powerful and prestigious organisation (Smith & Freyd, 2014). Freyd has now collaborated with many researchers to produce an impressive body of work on institutional courage, an antidote to organisational betrayal. These two concepts also have broader applications and relate well to any organisation using the tactics described in this book.

Organisational betrayal largely stems from organisational leaders who display negative emotions such as hubris, but also cowardice. Organisational expressions of cowardice and self-protection become ingrained in organisational systems, necessitating the development of institutional courage to displace them. The evolution of institutional courage can be traced through several important papers (Freyd, 1997; Harsey & Freyd, 2020; Smidt et al., 2023; Smith & Freyd, 2014). Institutional courage refers to proactive, transparent, accountable behaviours by organisations and their leaders when addressing harm. While betrayal involves halo protection, paltering, or scapegoating, institutional courage involves truthful acknowledgement, robust legal accountability, and flexing of moral fibre.

Institutional courage is not a vague aspirational quality but a series of measurable behaviours that have measurable outcomes. A large-scale study of employees who had experienced workplace sexual harassment found that institutional courage was strongly associated with better psychological and physical health, higher job satisfaction, and more substantial organisational commitment (Smidt et al., 2023).

It is evident that institutional courage transcends the performative virtue contained in meal-room values statements, because it acts with benevolence, truthfulness, and curiosity.

Practices of Institutional Courage

In a 2018 article, Freyd provided ten practical steps that display courage (Freyd, 2018). While her list directly responds to allegations of sexual assault, the general wisdom is useful across all workplace settings. The list extends from obvious actions, such as obeying the law, to broader ones requiring organisational leaders to make brave decisions against corporate culture.

Another way of thinking about institutional courage is the challenge to turn organisations from corporate thermometers into cultural thermostats. Thermometers adjust to the prevailing environment by representing the cultural temperature in conformity to that culture. Thermostats set a new cultural temperature and bring others up to that mark. Thermostatic organisations lead, while thermometric organisations follow.

The first action Freyd gives in the process of becoming a thermostatic organisation is to obey, and even transcend, workplace safety codes, criminal laws, and civil rights. An approach of compliance that defends senior leaders and organisational reputation fuels cynicism and embitterment among staff. Cynicism is fuelled when a senior leader does a safety walk to check compliance and 'consults' staff about changes they've already made. When those leaders also speak of organisational obedience to laws as a magnanimous gesture, workers smell the paltering and become embittered.

Therefore, press releases should not palter by implying that obeying the law is a display of corporate virtue. Instead, collaboratively designed systems should aim to make workplace systems morally laudable, not just legally defensible. Legal departments should not simply defend the organisation but also the rights of those threatened with punitive compliance, which is often administrative violence. Leaders must ask if the legal team have the mandate and will to tell the CEO the organisation is sailing close to the legal and moral wind when it harasses whistle-blowers. Further, does the CEO have the courage to change tack when challenged? To make your system laudable and brave, the following practical steps should be implemented.

Overtly, move organisational inquiry systems from an adversarial mindset to an inquisitorial approach. In criminal courts, the adversarial model pits two parties against each other as a way of determining guilt and punishment. In organisations, individuals are targeted for investigation, placing that person under threat of punishment. Three main questions are asked: what rule was broken, by whom, and how should they be punished?

Alternatively, an inquisitorial approach, as seen in coroner's courts, seeks not to assign blame but rather to determine: what happened, why it happened, and what systemic and individual changes might prevent recurrence. The emphasis is on understanding rather than blaming, making participants more willing to contribute and allowing for identification of systemic contributors to the event.

An inquisitorial approach is a better model for addressing workplace issues such as whistleblowing, error, or failure, as it prioritises truth over blame, fosters psychological safety, and produces practical recommendations that address procedures, incentives, training, and culture to reduce future risks. Inquisitorial learning should have clearly articulated thresholds for when an adversarial or disciplinary process is warranted, such as in deliberate misconduct, fraud, or repeated negligence. In practice, this means investigating first in a coroner-style mode to establish facts and root causes, and then applying proportionate consequences if sufficient evidence supports them.

It is important to pause and note that organisational leaders are under pressure from government regulators. Many regulators boast about the fines imposed on rogue operators, often at the expense of communicating the lessons learnt. Some years ago, I confronted a regulator with this idea, and a senior leader replied, 'We balance the carrot with the stick'. However, I was able to point out that 19 of the 22 news stories on their website highlighted punishments meted out, and none highlighted lessons learnt from investigations. The leader's head fell, and he could only say, 'point taken'. Regulators and employers will improve safety by highlighting learning. Once systemic learnings have been achieved, they should be communicated widely throughout the organisation.

Another practical legal step to obtain organisational courage is to stop the overuse of non-disclosure agreements (NDAs). NDAs have their place in protecting intellectual property, commercially sensitive material, and innocent individuals. However, they are also routinely used to mask pervasive and systemic harm to workers and silence complainants (Victorian Equal Opportunity & Human Rights Commission, 2022). A large church body in my city also recently admitted that 'NDAs in [some] circumstances can be used to cover up ungodly behaviour or to protect an individual or organisation from reputational damage' (Sydney Diocese Standing Committee, 2024).

The routine use of NDAs fosters a culture in which leaders maintain the presumption that their actions are secret and therefore largely unaccountable. Conversely, whistleblowers and scapegoated employees often sign NDAs due to financial pressures resulting from job loss and to prevent threats to their reputation. Organisational reputational risk is a strong motivation for the overuse of NDAs, and a cowardly way to deal with dissent.

To prevent their abuse, organisations should provide a publicly available statement on the number and category of NDAs struck each year, such as commercial-in-confidence, whistleblower settlement, and the like. This should be measured against a benchmark figure for NDA usage that is agreed upon by all stakeholders. Such a register serves as a robust metric for stakeholders about an organisation's commitment to transparency and truthfulness.

Finally, organisations and institutions should appoint an ethics committee to oversee whistleblowing matters, prolonged investigations, sanctions against workers, or matters of ethical and moral significance. The use of an ethics committee is part of the practice that Freyd refers to as engagement in self-study. Ethics committees are widely used in the research world to ensure that no harm is done to participants;

yet, punitive workplace structures rarely, if ever, consider the damage caused to those being investigated.

Risks and benefits to those who are being investigated should be considered. For example, when investigations are prolonged or transgress model litigant standards, ethics committees could compel accountability and rectification of unfair practices. Committees should be tailored to each industry and comprised of individuals external to the agency who possess a diverse range of expertise, including legal, human resources, employee representatives, community representatives, psychology, and, importantly, pastoral care.

These practices help rewire the organisation's defensive instincts where an intuitive DARVO response has been operating. Without these safeguards, victims have their credibility pre-emptively and systemically attacked, and perpetrators are emboldened to betray others through opaque systems (Harsey & Freyd, 2020). On the positive side, institutions that enact these processes will experience that public trust is maintained by transparency and contrition rather than fines and shame arising from a *mea culpa*.

Organisations that implement institutional courage as a proactive governance principle, rather than a crisis response, signal to stakeholders that their moral fibre is as important as their organisational reputation, operational performance, or financial outcomes. In crises, such organisations have reserves of trust, and in times of calm, they foster cultures of fairness and psychological safety.

DE-RISKING THE WORKPLACE

I attended a conference some time ago, and although my memory of the presenter is dim, the image she presented has remained with me. With an image of a massive circular saw blade behind her, the presenter asked us to imagine a sawmill that used large saws to cut timber had an inspection to check their safety processes. Imagine then that the mill operators tell you about their saw blade resilience classes, and their regular morning teas to discuss saw blade injuries.

You ask the owners if there are automated systems, physical barriers, or shutdown mechanisms in place to protect workers. Then imagine they said they did not need them because they had seen blade resilience classes and a great trauma hospital nearby that could reattach any severed body parts. You might demand that physical action be taken to eliminate the risks associated with operating such hazardous equipment and ensure that it meets its Work Health and Safety (WHS) requirements.

The physical safety of workers has been highly regulated for many years in most advanced economies, and WHS regulators have sharp teeth to ensure workers' safety. Until recently, the main object of WHS regulators has been to encourage and enforce the mitigation of physical workplace safety risks. Most people are aware of the dangers of trip hazards, electricity, falling from heights, or even saw blades. Still, until recently, many workplaces have treated psychological injuries, such as the negligent sawmill operator above, using resilience workshops and post-event first aid as solutions. Recently, there have been significant changes in international standards that recognise the need to prevent psychological injuries arising in the workplace.

This change to consider psychological mechanisms of harm is based on years of research about psychosocial safety climate in organisations (Dollard et al., 2012; Dollard & Bakker, 2010). This work culminated in the creation of an international standard, ISO45003, in 2021 (International Organization for Standardization, 2021). Now organisations are increasingly compelled to recognise and mitigate the risk of psychological harm from more than a dozen hazards that include these four key areas: poor supervisor support, work overload, poor procedural justice, and low reward and recognition (see Chapter 3).

These highlight the connection between psychosocial hazards and betrayal-type moral injury. The ISO45003 was released during my doctoral research. Just before its release, I had analysed interviews with my research participants, and I selected the central themes associated with betrayal and abandonment. To my surprise, these four key themes aligned highly with the psychosocial hazard categories. That is, psychosocial hazard categories do an outstanding job of operationalising the kinds of behaviours people experience as betrayal.

As a result, psychosocial risk mitigation is increasingly being mandated and is a relevant way to prevent betrayal and moral injury. When I noticed the connection between moral injury and psychosocial risk, it helped make sense of a prescient article by ethicist Dr Matthew Beard (Beard, 2018). Three years before ISO45003, he applied a WHS analogy to moral injury, noting it should be treated as a workplace injury because organisations can cause moral harm. My data confirmed that psychosocial hazard exposure can lead to moral kinds of injuries. Therefore, employers must not just think of moral injury as a wellbeing/mental health issue, but also mitigate the possibility of it arising through WHS frameworks.

Practices of De-risking

Organisations should rapidly implement a range of actions to mitigate psychosocial risks. Many consultants are well-experienced in providing specific advice, so I will leave this to those working consistently in that field. Instead, I suggest some concrete and thermostatic actions below that can create cultures where changes can be implemented.

The first is to recognise that moral injuries are different to physical injuries. Moral injuries are more costly and impact a person's character, sense of worth, and meaning in more profound ways. It is well known that serious mental health claims cost around five times more to resolve than all injury claims, and recovery time is five times longer (Safe Work Australia, 2018). Also, when physical injuries are associated with a perceived injustice, they take longer to heal, again driving up costs and harm (Sullivan et al., 2008).

To create an environment where psychosocial risks are effectively addressed, several key principles are essential. The first is to recognise that no department owns the mandate to address these issues. Organisational 'knife fights' between departments over funding and prestige that comes with the responsibility to manage this risk are not acceptable. C-suite leaders must implement systems to encourage their Risk and Safety personnel to work collaboratively with People and Culture staff, particularly staff wellbeing workers.

Included in this collaborative mix must be the legal and investigative sections of your workforce who are heavily implicated in the infliction of workplace betrayal through administrative violence. The legal department may be one of the most significant psychosocial hazards in the workplace, but it can also be its biggest protector. As a priority, senior leaders are to utilise their hostage-negotiating skills, including empathy and calibrated questions, with departmental leaders to help them collaborate with other departments.

Secondly, do not come up with solutions at the management level. Key to mitigating psychosocial risks is the need to collaborate with workers in co-design and review. Too often, leaders claim they have consulted staff, by which they mean leaders have imagined what the problems are, created solutions, and then given staff one week to provide feedback, which will be largely ignored. Staff who have been 'consul-told' like this have little buy-in.

Instead of consul-telling staff about their problems and the solutions key leaders have developed, it is important to understand staff and how they experience those issues. Again, the ingredients of empathy, benevolence, and civility are key to collaborative design and review. Approach staff and ask about the differences they encounter between their position description and standard operating guidelines, as well as their experiences with the reality of day-to-day operations. Ask staff what the real difficulties are and seek to minimise the differences between what executives imagine and the actual work done by technicians.

'Sometimes the biggest problem is not the mountain one must climb, it is the pebble in one's shoe'. By all means, ask your staff about their pebbles, but even better, walk a mile in their shoes. It is essential for those charged with psychosocial risk management to put aside some of their schedule to be with staff and experience their problems firsthand. When a solution has been collaboratively identified and implemented, it must continually be reviewed and approved. When a date for review is put on a document, review it with the staff who helped create it. This gets staff buy-in and reduces the sense that managers are the enemy to be vanquished.

Finally, senior leaders must create the conditions for psychological safety, a term that is often confused with similar ones. Therefore, educate senior and middle managers about the difference between psychological safety, psychological health and safety, psychosocial hazards, and psychosocial safety climate. Admittedly, adding some variation to the adjectival use of 'psychology' might help reduce confusion; however, getting help from experts will build the capability and capacity to address them all.

Psychological safety is a cultural element that helps foster psychological health and safety, as well as mitigate psychosocial risk. In essence, it means your workers experience a place where they can say 'that is not right', or 'I got it wrong before', without being ridiculed, excluded, scapegoated, demoted, or fired. I witnessed this firsthand when a medical practitioner admitted to colleagues that they could not successfully perform a lifesaving procedure; however, their actions luckily had a different positive effect. The practitioner admitted that they saved a life by dumb luck, not skill. Clinicians new to that workplace were stunned they would admit this, but appreciated the admission and learnt valuable, and possibly life-saving, lessons.

Senior leaders must set the pace by embracing psychological safety, which involves admitting their own failures and thereby emboldening their staff to do the same. Further, people who turn failure into a learning opportunity should be valued and given a greater voice. This can only occur when organisations are curious, have an inquisitive approach, and collaborate on solutions. When these ingredients are in place, your organisation will be able to build strong systems to mitigate risks associated with staff betrayal, have a safer and more productive workplace, and be less likely to face public apologies.

HELP THEM TO LEAD: SHEPHERD LEADERSHIP

Mid- to upper-level managers are many things. They are the interface between senior leaders and frontline workers, meaning that they are the meat in the sandwich between the plans of senior leaders and the implementation of those plans by the workforce. Middle managers don't get to make decisions; they must enforce them. Because they are vulnerable, they often become perfect scapegoats if things go wrong. They are also stuck between their own hopes for higher leadership and the realities of their current role.

I regularly survey middle managers during my consultations with organisations and ask them why they chose a leadership role. They regularly claim that they entered leadership to help workers and advocate for them, yet often find themselves enforcing unrealistic work schedules and undertaking bureaucratic tasks, such as reporting on key performance indicators. Many leaders feel let down by the systems in which they hold leadership positions, resulting in many experiencing their own latent moral injuries. Yet they are expected to be all things to all people.

Middle managers must have some knowledge of the industry they work in, as well as the ability to manage finances, roster staff, and comply with various regulations unrelated to their training. Mental health is a new field in which expertise is expected, and many promote the need for leaders to be trained in mental health awareness. While these programmes can be well-meaning, there are mixed results regarding their success, and they add more pressure to managers.

I believe many mental health training packages for managers are well-meaning but misdirected. A better direction than helping them recognise mental health issues is to help managers be the leaders they always wanted to be. This approach has the added benefit of directly addressing the psychosocial hazard of 'poor leadership support'. During my research, I developed a model of shepherd leadership that gave leaders a solid identity and nine actionable activities to fulfil it.

Habitus and Habits of Shepherd Leaders

Australia was once known as the country that rode on the sheep's back. Australia has always had a strong wool and lamb industry that has been a mainstay of its economy and the stuff of folklore. Our pseudo-national anthem, Waltzing Matilda, is about a sheep thief! In Australia, sheep farmers often use motorbikes and sheepdogs to push the flock from behind when moving them to good pasture. However, in ancient times and different lands, shepherds walked in front while sheep willingly followed, a situation human leaders aspire to.

The ability to walk confidently into battles and workplace projects with the knowledge that your staff will follow you demands that the shepherd leader know their sheep, and the sheep trust the shepherd. Shepherds in this model are not strictly masters of their flocks but servants – protecting, providing, and healing those in their flock. Shepherd leadership works with the motives many new leaders already have to serve those they lead. Shepherd leaders create conditions where followers willingly follow because of the high levels of trust that have been created.

While many mistakenly believe that power and coercion are central to the exercise of leadership, trust is the currency of leadership. Trust, unlike real money, cannot be extorted; it can only be earned, and when earned, is also given willingly. Each of the actions below is part of a leadership training module under development, which I am rolling out to organisations, and will be a follow-up manual to this book. Shepherd leaders are marked by nine attributes, the second of which is the only one I will expand on here.

1. **They lead**: without abdicating the responsibility and authority given to them, they take a moral view of their authority, holding all titles of leadership lightly through a firm, fair, and friendly way of acting. They recognise they have been given power to lead, and use it wisely.
2. **Ethos of service**: good leadership is lifting others up. This is the essence of cultural and leadership change that must be realised in organisations. The ethos sees that power is not given *over* to people lower in the hierarchy, but for the service of those people. This inversion is not a weakness and requires greater strength, courage, and mutual trust between leaders and followers. The ethos of service turns the entire organisation on its head. The two figures below demonstrate this principle.

Traditional organisational charts (Figure 10.1) inculcate a culture of 'power over' others that demands trust and obedience from those lower in the system as a matter of duty. These power-driven systems can facilitate abusive leadership. These systems are often siloed (indicated by the dotted lines) with each directorate pitted against each for scarce power and funding. It pushes down on those with the least power who are also at the frontline of providing the service that the organisation exists for. Such a system places leadership over operational outcomes, stifles psychological safety, and embitters staff.

An inverted structure still understands the importance of power but uses it for those whose work is central to the organisation. Figure 10.2 depicts a system of 'power for' those who lack power but produce frontline organisational outcomes. This system builds trust and collaboration across the workforce and lifts the core work of the organisation above those who lead. To invert

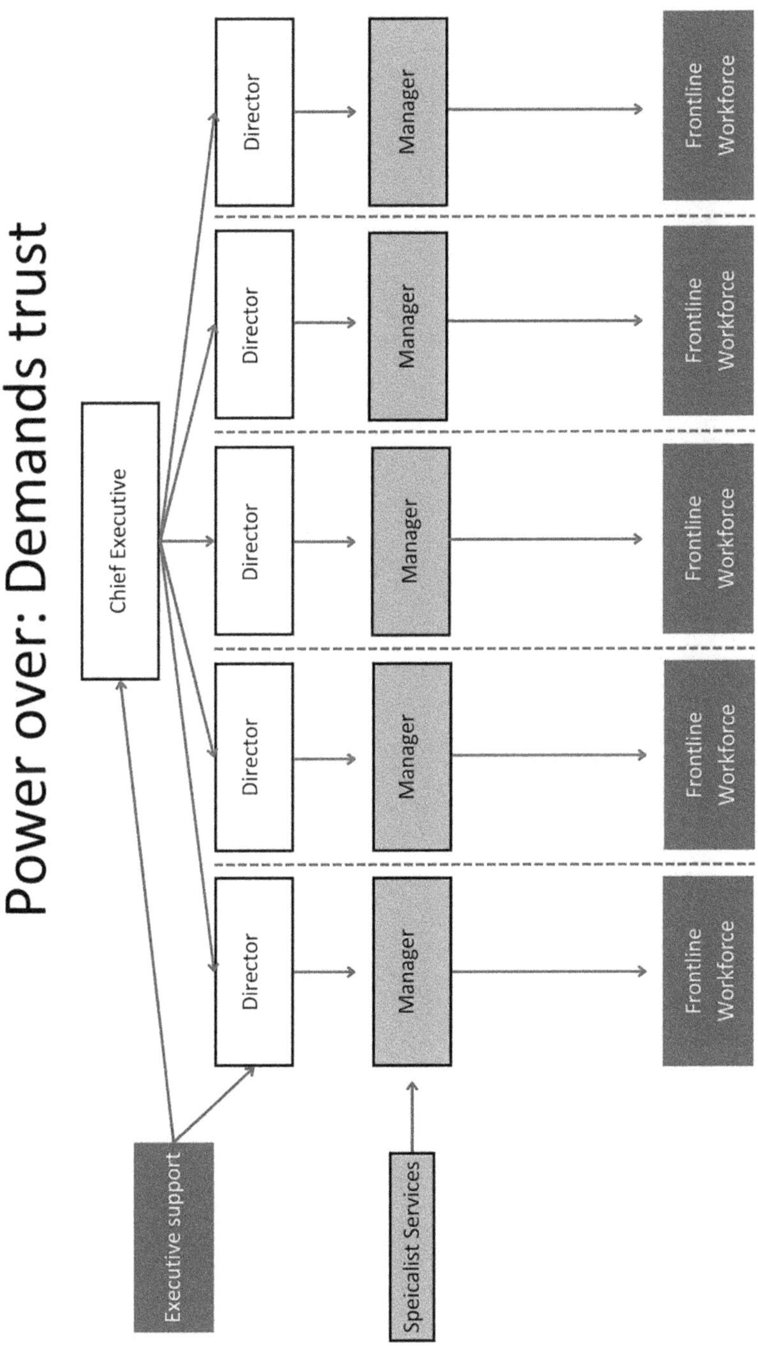

FIGURE 10.1 Traditional 'power over' organisational chart.

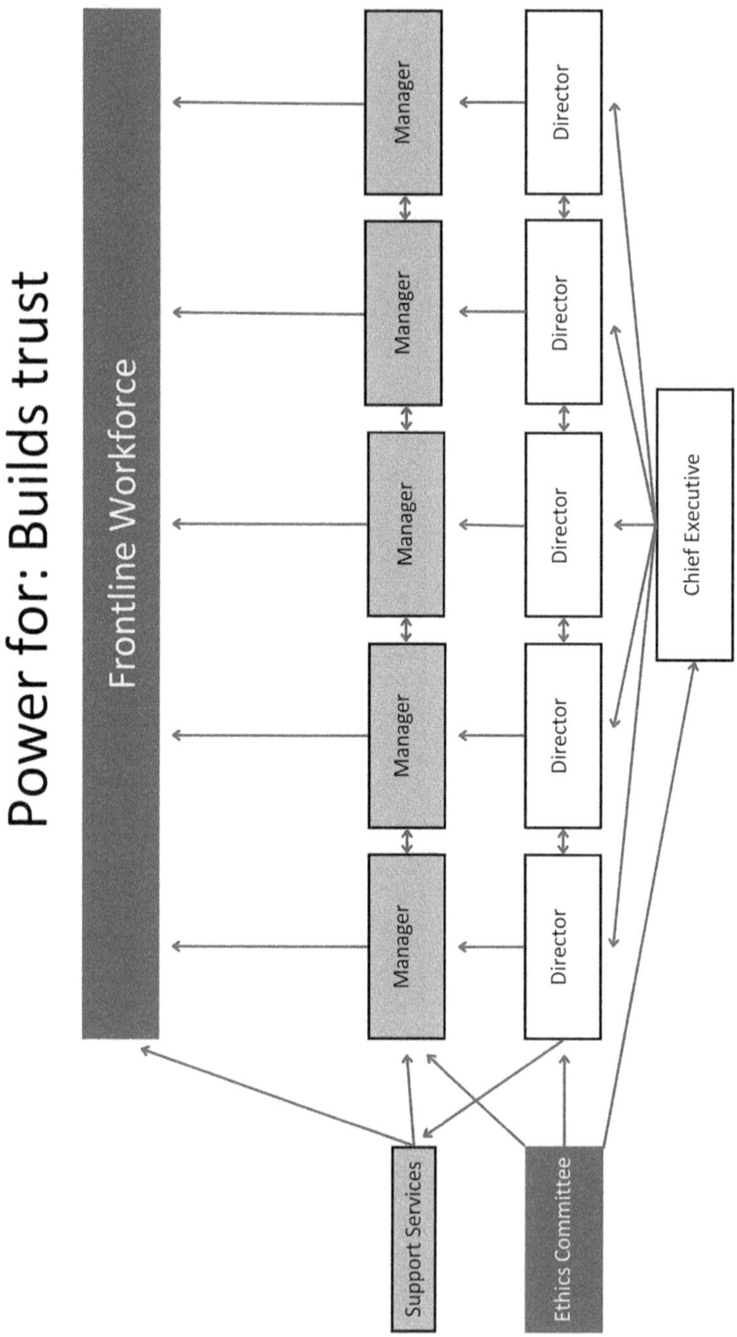

FIGURE 10.2 Inverted 'power for' organisational chart.

this structure requires courage and humility that a true leader is not afraid to embrace. Benefits of 'power for' systems are that it empowers collaboration, nurtures psychological safety, reduces psychosocial risk, and promotes a healthy and well workforce. If leaders want to build trust and accountability in their systems, they must invert the organisational hierarchy.

3. **Presence with staff**: in preference to scheduled meetings, shepherd leaders prioritise the genuine connection of the ad hoc hallway conversation and create opportunities to walk in the shoes of their staff regularly.
4. **Four-dimensional communication**: see Chapter 8.
5. **Provides**: ensures staff have the time and resources they need to do their work safely.
6. **Protects**: in acts of benevolence, leaders stand between their staff and the internal (company politics) and external (operational and emotional) dangers they face.
7. **Grateful**: transcending the mere expression of thanks, gratefulness flows from a state of gratitude which speaks positively of team members to self and higher leaders.
8. **Follows**: leaders submit themselves to another authority or leader so that they continually habituate and exemplify the experience of followership.
9. **Restores**: seeks to learn from adverse incidents and restore those who have fallen so they can be socially connected and vocationally active.

Together, these practices embody an organisational culture driven by a leadership style capable of building trust, restoring agency, and nurturing resilience in those harmed by unsupportive systems. These all feed into what is possibly the most developed systems approach to safety and wellbeing I have encountered, *Just Cultures*.

SYSTEMS APPROACH: CREATING A JUST CULTURE

How does one solve systemic procedural injustice issues in an organisation? One suggestion provided by a safety regulator is to display posters with aspirational cultural values in prominent areas of the workplace. Like other meal room posters, these ethics posters will be just visual noise that is turned to zero within a week. The suggestion to create posters of desired cultures to mitigate injustice is equal parts naïve and impotent.

Serious cultural issues deserve grown-up solutions, and the Just Culture framework is a sophisticated system. A key leader in Just Culture research is Sidney Dekker, whom we met in Chapter 7. Dekker's interest in Just Cultures is multifaceted: as a pilot, it is central to aviation safety; as a chaplain, he cares about people and their restoration; and as a professor, he can report that it positively impacts culture. Professor Dekker has written extensively on Just Cultures, and I recommend his work be read in full. However, it is helpful to outline three key areas for action at the conclusion of this book, as demanded by Just Culture research.

Practices of Just Cultures

These practices fall predominantly into three main areas. First, substantive justice addresses whether regulations, rules, and procedures are fair and legitimate. Second, procedural justice demands legitimate processes for determining any breaches of substantively fair procedures. This approach includes protections for both wrongdoers and whistleblowers and determines who should make such determinations. Third, restorative justice principles aim to restore the status of wrongdoers and heal relationships and injuries, moral and otherwise, of victims and survivors of ethical breaches (Dekker & Breakey, 2016).

Substantive Justice

French philosopher Anatole France famously noted something strange about the law: 'The law, in its majestic equality, forbids rich and poor alike to sleep under bridges, to beg in the streets, and to steal their bread'. In essence, the rich will never fall foul of this law, while the poor and powerless have little option but to break it. The same law can be unfair because it has vastly different impacts on people. In reflecting on these words, a senior judge noted 'traditional principle of Aristotelian ethics — that injustice inheres as much in treating unequals the same, as it does in treating equals differently' (The Honourable Spigelman AC, 2006).

Substantive justice in the workplace requires that legitimate interests and prior knowledge of staff are taken into account when rules are made – they should not unfairly impact them nor be outside the range of enforcement for senior leaders. To this, Dekker and Breakey (2016, p. 190) write substantive injustice is possible, 'where the larger institution and its executives develop safety standards precisely to absolve their legal blame for accidents by ensuring the availability of lower-level scapegoats who are bound to have failed one rule or another'.

Substantive justice demands that various principles of the rule of law must apply equally to corporations and to governments. Rule of law principles include that laws are known by staff, that there are strong checks and balances on those who write and enforce those rules, that everyone is subject to them, and that everyone receives fair treatment (Rule of Law Education Centre, n.d.).

The introduction of an externally staffed ethics committee is an important check and balance on the power of organisational leaders. Cases such as the scapegoating of managers in the Robodebt and 737MAX scandals show that rules were not equally applied to all those who were culpable. Furthermore, bureaucratic systems that produce an excessively large number or long sets of rules fail to make work safer or fairer. Such excessive rule making is the process of making work 'paper safe' to cover organisational liability rather than providing operational safety to workers (Smith, 2018).

Therefore, to create workplace substantive justice, it is crucial that rules about central issues such as human error, harassment, abuse, and whistleblowing are made in collaboration with impacted staff. Rules must not be in lengthy legalese and must be enforceable on senior leaders with the same proportion as they are for middle managers and frontline workers.

Procedural Justice

Second, procedural justice demands that these substantively fair rules are acted upon in a fair way. The use of administrative violence, often employed against whistleblowers, constitutes a direct breach of procedural justice. As a foundational principle, those who come forward with concerns, complaints, or admissions of error should be treated with kindness, courtesy, and understanding. These principles are important throughout the process of the administration of organisational justice. Apply benevolence, civility, and empathy before, during, and after any declarations of complaint or error.

Prior to allegations being brought, organisations must develop a culture where psychological safety is the norm and there is no need to call on personal heroics to tell the truth. The clinician should not have to be a hero to admit error, the victim of abuse should not have to be a hero to give their account, and the likes of Mike should not have to resort to heroics to protect vulnerable staff. Such actions should be normalised, and those who come forward valued, because elevating moral behaviour to an act of heroism shows that the entire culture has had a systemic moral failure (Dekker et al., 2022).

When an issue is raised, leaders must have been trained AND emboldened to take an inquisitorial rather than adversarial tone. When a whistleblower is belittled, attacked, or their competence or sanity questioned, it becomes a form of gaslighting that is psychologically and existentially harmful. Ensure that the phrases of hubris named in Chapter 5 are known by leaders and avoided. Avoid using trite euphemisms or clichés such as 'if this makes you feel better', as these belittle and demean people. Instead ensure those who are trained to receive complaints know trauma informed principles and can respond with tactical empathy and active listening.

During any investigation, the questions asked should be those that address systems issues as a priority over an individual action. If an allegation arises against or by a staff member 'Bob', it would be all too easy to locate the investigation on the person by asking, 'Why did Bob act in such a way, what was he thinking?' Instead, use calibrated questions such as, 'What are Bob's thoughts (reasons) and feelings (emotions) about this issue right now?' This question gauges their willingness to be open and honest.

Ask also, 'How could Bob's decision be seen as legitimate at the time it was made?' This question does not presume negligence or guilt but seeks to understand how systems may have failed to predict the situation faced at the time. Ask the questions that are really needed: 'What system might have prevented this from occurring, and what will ensure that it does not happen again?'

Finally, to ensure procedural justice during an investigation, other principles of the rule of law should be applied. It is favourable to have external oversight of the inquiry to ensure conflicts of interest do not arise and to ensure that organisational reputation is not favoured over truth-seeking and telling. Proper avenues of appeal and support must be provided to those undergoing investigation either in whistleblowing or cases of alleged error or malice. When the investigation is completed, there should be a preference to restore the whistleblower to full community in the workplace. Likewise, those who have not maliciously committed an offence, but instead made oversights or errors should be restored as a preference.

Restorative Justice

Legend tells of the former CEO of IBM, Tom Watson Sr, who was a fallible but honourable man with deep convictions about restoration. Under his tenure, IBM had just survived the Great Depression and prevented widespread layoffs through Watson's grace and tenacity. However, these were wearing thin when an important tender arose – it was a sink or swim proposition. Unfortunately, a key staff member tasked with the tender failed badly and IBM's future was placed in doubt.

The story goes that the staff member drafted his resignation and placed it on Watson's desk. Instead of accepting it, Watson was curious and gave the staff member an opportunity to give his own account. After showing a level of self-awareness of his failure, Watson refused his resignation, instead looking him in the eye and saying, 'Why would I accept this when I have just invested one million dollars in your education?'

Restoring others is not weakness – it may be kindness, but can also be sensible business practice. Every wrong turn taken is one that need not be travelled again, and the person travelling the wrong road is the best to warn others of dangers. Removing staff for error or raising concerns is the real wrong turn. Routinely firing staff for errors shuts down open communication about near misses. Scapegoating whistleblowers creates a dangerous culture of suspicion and deviance. The hope of restoration brings openness and healing.

A key way to encourage restoration is through a change in the way accountability is understood. Firstly, it moves beyond the backward-looking notion of finding the rule broken, the rule breaker, and then punishing them and passing it off as accountability. Instead, a forward-looking accountability asks how your organisation will be better tomorrow as a result of lessons learnt today, seeing staff as the solution to be prospectively harnessed, not a problem to be retrospectively controlled (Dekker & Breakey, 2016). Restoration is key to the creation of moral fibre.

We have a problem in our world in which error and suffering are seen to be avoided at all costs, and moved on from as soon as possible. However, they are our greatest teachers, if only we will take the time to listen. Making an expensive or fatal error or being victimised as a whistleblower is a terrible experience, and we have discussed the danger of moral degradation. However, restoration opens the door to growth. Therefore, leaders must develop a publicly available organisational position statement on restoration and forgiveness.

FORGIVENESS

In February 2020, a terrible motor vehicle collision occurred near my home. Four young children – three siblings and a cousin – were run over and killed by a speeding, drug-affected driver who was arrested at the scene and imprisoned for many years. In this dark moment, a moment of bright light was seen. Before any sentencing occurred, before any statement of sorrow by the driver, Leil Abdallah, mother of those three siblings, publicly declared forgiveness for the driver. The children's father, Danny, later visited the driver in prison, an act that helped bring healing to both men.

The media was unsure of how to respond when this proactive strike of forgiveness occurred. Was this act of forgiveness naïve, courageous, unwise, or loving? I believe it was an audacious act of hope, mercy, strength, and healing. Danny Abdallah later said 'Forgiveness is like… a muscle. You need to keep building it'. I like this idea that forgiveness requires the strength that comes from exercising one's moral fibre. Forgiveness is not capitulation to evil; it confronts it and requires a moral muscularity many lack.

Forgiveness does not mean that temporal punishment is to be negated; that speeding driver is still appropriately in prison. Forgiveness does not mean reconciliation has occurred, as fault may be denied, or repentance and recompense refused by the offending party. Forgiveness cannot be demanded or forced; it is only real when it is a free choice of the heart, soul, and mind.

Forgiveness discharges any personal debt that is owed, which can be costly. It forgoes one's right to vengeance, stops retaliation, and even refuses to wish the evil done to the forgiver be done to the offender. Forgiveness halts the downward trajectory of moral degradation and frees both the forgiver and the forgiven. While betrayal lives in the shadow of a halo, it dies in the light of forgiveness.

This book begins with the story of Mike, who was deeply hurt by betrayal and was tempted towards vengeance. Many other accounts have been presented in these pages, and all have left open the possibility of revenge, as well as that of forgiveness. When you consider your pain and suffering, the mention of forgiveness may seem almost offensive. However, forgiveness is the most profound power we have against betrayal and moral injury. One of my key hopes in writing this book is that organisations would adopt forgiveness as a key value, and individuals embrace it as a core character trait.

Forgiveness is costly, and it is important to weigh that cost as you consider forgiving those who have betrayed your trust and caused deep moral injuries. However, people rarely calculate the cost of resentment, which is akin to drinking poison and hoping that their enemy gets sick. Those who fail to forgive die a slow inner death resulting from the poison of resentment. That resentment not only destroys you but also destroys relationships with those that you hold dear, displaces across relationships and roles, prevents the construction of trust with others, and makes any job a prison of doubt and cynicism.

Nelson Mandela, the former political prisoner and South African president, valued forgiveness and reconciliation, and oversaw the transition from apartheid to full democracy. While many feared mass violence and hatred, he oversaw truth-telling, minimised vengeance, and brought peace. It is he who said, 'In the end, reconciliation is a spiritual process, which requires more than just a legal framework. It has to happen in the hearts and minds of people' (Mandela, 2012).

Indeed, Tolstoy's inner revolution centred on the discovery and enactment of forgiveness. If forgiving a betrayer is too large a revolution for you right now, may I suggest that you do something that requires even more bravery – seek forgiveness for the wrongs you have done. At some point in your life, you have failed, for to err is human. And in your erring, you have hurt and possibly betrayed others.

If forgiveness is not possible for you right now, seek and find forgiveness from the ones you have hurt. It may be that, in the light of your own forgiveness, you can forgive others. As much as any time in history, our world needs the forgiveness that stops hatred, vengeance, resentment, war, and betrayal. It will take personal courage and the support of others, as well as the humility of having experienced forgiveness, to unleash the revolution of forgiveness that is the antidote to betrayal. Forgiveness prevents a person from being crushed by betrayal and will instead diminish any power betrayal holds. May the reader receive and give forgiveness, and may the pain of your betrayal be vanquished along the way.

ITEMS FOR CONSIDERATION

- Which of the four frameworks – institutional courage, de-risking, shepherding, or Just Cultures – is most needed in your current workplace? Which do you have the power to start enacting or to contribute to?
- If you could pick one leadership behaviour to work on the most, how would you start to do that?
- What is the cost of the resentments you are currently holding? Who is your resentment hurting the most right now?
- Who can you forgive, and who should you seek forgiveness from?

References

ADF. (2021). *Australian Defence Glossary*. Canberra: Australian Defence Force.

Akstinaite, V., Robinson, G., & Sadler-Smith, E. (2020). Linguistic markers of CEO hubris. *Journal of Business Ethics*, *167*(4), 687–705. http://search.ebscohost.com/login.aspx?direct=true&db=eoah&AN=50162176&site=ehost-live

Altman, D. G. (2012). Building a metaphor: Another brick in the wall? *BMJ*, *345*(dec17 6), e8302–e8302. https://doi.org/10.1136/bmj.e8302

Archer, G. L. (1999). Covenant. In W. A. Elwell (Ed.), *Evangelical dictionary of theology* (pp. 276–278). Baker Books.

Atuel, H. R., Barr, N., Jones, E., Greenberg, N., Williamson, V., Schumacher, M. R., Vermetten, E., Jetly, R., & Castro, C. A. (2021). Understanding moral injury from a character domain perspective. *Journal of Theoretical and Philosophical Psychology*, *41*(3), 155–173. https://doi.org/10.1037/teo0000161

Auditor-General of NSW. (2025). *The mental health and wellbeing of NSW police: NSW Auditor-General's report*. www.parliament.nsw.gov.au/tp/files/191152/FINAL%20-%20The%20mental%20health%20and%20wellbeing%20of%20NSW%20police.pdf

Australian Army Research Centre. (2022). Recruitment for the future. *Australian Army Research Centre*. Retrieved August 21 2024, from https://researchcentre.army.gov.au/library/land-power-forum/recruitment-future

Baumeister, R. F., & Landau, M. J. (2018). Finding the meaning of meaning: Emerging insights on four Grand Questions. *Review of General Psychology*, *22*(1), 1–10. https://doi.org/10.1037/gpr0000145

Bayani, B. A. (2021). *Exploring the relationship between moral injury, supervisor leadership style, and suicidal behavior in Texas paramedics* [Doctor of Philosophy, Dallas Baptist University].

Beard, M. (2018). Moral injury is a new test for employers. *Ethics Alliance*. Retrieved 2 December 2022, from https://ethics.org.au/moral-injury-is-a-new-test-for-employers/

Berlinger, N. (2013). Writing wrongs: On narratives of moral distress. *Narrative Inquiry in Bioethics*, *3*(2), 131–137. https://doi.org/10.1353/nib.2013.0043

Bertrand, M., & Mullainathan, S. (2003). *Are Emily and Greg more employable than Lakisha and Jamal? A field experiment on labor market discrimination*. National Bureau of Economic Research. https://dx.doi.org/10.3386/w9873

Blumberg, D. M., Papazoglou, K., & Schlosser, M. D. (2022). *The POWER manual: A step-by-step guide to improving police officer wellness, ethics, and resilience*. American Psychological Association.

Bonhoeffer, D. (1954). *Life together*. SCM.

Bradford, K. (2023, January 20). Relating soul to spirit. *Pastoral Thinking*. https://pastoralthinking.org/2023/01/20/relating-soul-to-spirit/

Brown, B. (2012). *Listening to shame*. YouTube. www.youtube.com/watch?v=psN1DORYYV0&t=924s

Cameron, A. J. B. (2011). *Joined-up life: A Christian account of how ethics works*. Inter-Varsity Press.

Campbell, J. L., Darby, J., Oswald, P., Mackay, L., & Schofield, G. (2025). Shifting organizational sentiment: Qualitative insights from a mental ill-health prevention program for New Zealand firefighters. *Comprehensive Psychiatry*, *142*, 152615. https://doi.org/10.1016/j.comppsych.2025.152615

Chaplin, S. (1900). The stained glass political platform. *The Century Magazine, USA*, 304–309. https://unz.com/PDF/PERIODICAL/Century-1900jun/142-147/

Copenhaver, M. B. (2014). *Jesus is the question: The 307 questions Jesus asked and the 3 he answered*. Abingdon Press.

Cox, S. J. (2018). *An introduction to moral injury in defence*. Australian Defence College. Retrieved 23 July 2020 from https://theforge.defence.gov.au/publications/introduction-moral-injury-defence

Cox, T., Griffiths, A., Barlowe, C., Randall, R., Thomson, L., & Rial-Gonzalez, E. (2000). *Organisational interventions for work stress: A risk management approach* (1368-6828). (Health & Safety Executive, Issue. www.hse.gov.uk/research/crr_pdf/2000/crr00286a.pdf

Crosweller, M. (2022a). Disaster management and the need for a relational leadership framework founded upon compassion, care, and justice. *Climate Risk Management*, *35*, 100404. https://doi.org/10.1016/j.crm.2022.100404

Crosweller, M. (2022b). Disaster management leadership and the need for virtue, mindfulness, and practical wisdom. *Progress in Disaster Science*, *16*, 100248. https://doi.org/10.1016/j.pdisas.2022.100248

Crosweller, M. (2024). *Compassion in disaster management: The essential ethic of relational leadership*. Routledge.

Currier, J. M., Farnsworth, J. K., Drescher, K. D., McDermott, R. C., Sims, B. M., & Albright, D. L. (2018). Development and evaluation of the Expressions of Moral Injury Scale—Military version. *Clinical Psychology & Psychotherapy*, *25*(3), 474–488. https://doi.org/10.1002/cpp.2170

De Bruin Cardoso, I., Russell, A. R., Kaptein, M., & Meijs, L. (2023). How moral goodness drives unethical behavior: Empirical evidence for the NGO halo effect. *Nonprofit and Voluntary Sector Quarterly*. https://doi.org/10.1177/08997640231179751

Defazio, P. A., & Larsen, R. (2020). *The design, development and certification of the Boeing 737 MAX (Final Committee Report)*. Retrieved from https://democrats-transportation.house.gov/download/20200915-final-737-max-report-for-public-release

Dekker, S. (2007). *Just Culture: Balancing safety and accountability*. Ashgate.

Dekker, S. (2014). *The field guide to understanding 'Human Error'* (3rd ed.). CRC Press.

Dekker, S., Layson, M. D., & Woods, D. D. (2022). Repentance as rebuke: Betrayal and moral injury in safety engineering. *Science and Engineering Ethics*, *28*. https://doi.org/10.1007/s11948-022-00412-2

Dekker, S. W. A., & Breakey, H. (2016). 'Just Culture:' Improving safety by achieving substantive, procedural and restorative justice. *Safety Science*, *85*, 187. https://doi.org/10.1016/j.ssci.2016.01.018

Dobos, N. (2015). Moral trauma and moral degradation. In T. R. Frame (Ed.), *Moral injury: Unseen wounds in an age of barbarism*. UNSW Press.

Dollard, M. F., Bailey, T., McLinton, S., Richards, P., McTernan, W., Taylor, A., & Bond, S. (2012). *The Australian workplace barometer: Report on psychosocial safety climate and worker health in Australia*. Centre for Applied Psychological Research University of South Australia.

Dollard, M. F., & Bakker, A. B. (2010). Psychosocial safety climate as a precursor to conducive work environments, psychological health problems, and employee engagement. *Journal of Occupational and Organizational Psychology*, *83*(3), 579–599. https://doi.org/10.1348/096317909x470690

Drew, J. M., Bartels, S., & Herrington, V. (2023). *Leadership for wellness: A strategy for developing police and public safety leaders*. Leadership Development White Paper. A Collaboration between Griffith Criminology Institute Griffith University and AIPM.

Dunham, C. R. (2022). Social truths in the workplace: How adversarialism undermines discrimination litigation. *SSRN Electronic Journal*. https://dx.doi.org/10.2139/ssrn.4049012

Edelman Trust Institute. (2019). 2019 Edelman Trust Barometer: Global Report. www.edelman.com/sites/g/files/aatuss191/files/2019-02/2019_Edelman_Trust_Barometer_Global_Report.pdf

Edmondson, A. (1999). Psychological safety and learning behavior in work teams. *Administrative Science Quarterly*, *44*(2), 350–383. https://doi.org/10.2307/2666999

Elangovan, A., & Shapiro, D. L. (1998). Betrayal of trust in organizations. *Academy of Management Review*, *23*(3), 547–566. https://doi.org/10.2307/259294

Engel, G. L. (1977). The need for a new medical model: A challenge for biomedicine. *Science*, *196*(4286), 129–136. https://doi.org/10.1126/science.847460

Fani, N., Currier, J. M., Turner, M. D., Guelfo, A., Kloess, M., Jain, J., Mekawi, Y., Kuzyk, E., Hinrichs, R., Bradley, B., Powers, A., Stevens, J. S., Michopoulos, V., & Turner, J. A. (2021). Moral injury in civilians: Associations with trauma exposure, PTSD, and suicide behavior. *European Journal of Psychotraumatology*, *12*(1). https://doi.org/10.1080/20008198.2021.1965464

Fox, P. (2019). *Walking towards thunder*. Hachette Australia.

Freyd, J. J. (1997). Violations of power, adaptive blindness and betrayal trauma theory. *Feminism and Psychology*, *7*(1), 22–32. https://doi.org/10.1177/0959353597071004

Freyd, J. J. (2018). When sexual assault victims speak out, their institutions often betray them. *The Conversation*. https://theconversation.com/when-sexual-assault-victims-speak-out-their-institutions-often-betray-them-87050

Fuchs, J. R., Fuchs, J. W., & Mescher, M. (2023). The Ignatian Examen in medical practice: A model for clinicians. *Journal of Religion and Health*, *62*(4), 2341–2348. https://doi.org/10.1007/s10943-022-01636-z

Fuller, A. (2009). *Tricky people: How to deal with horrible types before they ruin your life*. Finch Publishing.

Gilmartin, K. (2002). *Emotional survival for law enforcement: A guide for officers and their families*. E-S Press.

Gilmartin, K. (2014). Keeping good people good. *Police News*, 17–21. https://emotionalsurvival.com/KEEPINGGOOGPEOPLEGOOD.pdf

Grant, A. (2023). *Think again: The power of knowing what you don't know*. Penguin.

Gray, K., & Wegner, D. M. (2011). To escape blame, don't be a hero—Be a victim. *Journal of Experimental Social Psychology*, *47*(2), 516–519. https://doi.org/10.1016/j.jesp.2010.12.012

Greenberg, N. (2022). *Moral injury, mental health and healthcare staff*. UKPTS Conference 2022, King's College London.

Haidt, J. (2013). *The righteous mind: Why good people are divided by politics and religion*. Vintage Books.

Harsey, S., & Freyd, J. J. (2020). Deny, attack, and reverse victim and offender (DARVO): What is the influence on perceived perpetrator and victim credibility? *Journal of Aggression, Maltreatment & Trauma*, *29*(8), 897–916. https://doi.org/10.1080/10926771.2020.1774695

Hillman, K. (2010). *Vital signs: Stories from intensive care*. NewSouth Publishing.

Hodgson, T. J., & Carey, L. B. (2017). Moral injury and definitional clarity: Betrayal, spirituality and the role of chaplains. *Journal of Religion and Health*, *56*(4), 1212–1228. https://doi.org/10.1007/s10943-017-0407-z

International Organization for Standardization. (2021). Occupational health and safety management — Psychological health and safety at work — Guidelines for managing psychosocial risks. In (Vol. ISO 45003:2021).

James, W. (1909). *A pluralistic universe: Hibbert Lectures at Manchester College on the present situation in philosophy* (Vol. 4). Harvard University Press.

Jamieson, N., Carey, L. B., Jamieson, A., & Maple, M. (2023). Examining the association between moral injury and suicidal behavior in military populations: A systematic review. *Journal of Religion and Health*, *62*(2023), 1–22. https://doi.org/10.1007/s10943-023-01885-6

Johnson, M. (1985). Imagination in moral judgment. *Philosophy and Phenomenological Research*, *46*(2), 265–280. https://doi.org/10.2307/2107356

Johnson, M. (1993). *Moral imagination: Implications of cognitive science for ethics*. University of Chicago Press.

Johnstone, M.-J., & Hutchinson, A. (2015). 'Moral distress' – time to abandon a flawed nursing construct? *Nursing Ethics*, *22*(1), 5–14. https://doi.org/10.1177/0969733013505312

Jones, N., Seddon, R., Fear, N. T., McAllister, P., Wessely, S., & Greenberg, N. (2012). Leadership, cohesion, morale, and the mental health of UK Armed Forces in Afghanistan. *Psychiatry: Interpersonal and Biological Processes*, *75*(1), 49–59. https://doi.org/10.1521/psyc.2012.75.1.49

Kelley, M. L., Chae, J. W., Bravo, A. J., Milam, A. L., Agha, E., Gaylord, S. A., Vinci, C., & Currier, J. M. (2021). Own soul's warning: Moral injury, suicidal ideation, and meaning in life. *Psychological Trauma: Theory, Research, Practice, and Policy*, *13*(7), 740–748. https://doi.org/10.1037/tra0001047

Kline, P., Rose, E. K., & Walters, C. R. (2022). Systemic discrimination among large US employers. *The Quarterly Journal of Economics*, *137*(4), 1963–2036.

Laker, B., Pereira, V., Malik, A., & Soga, L. (2022). Dear manager, you're holding too many meetings. *Harvard Business Review*. Retrieved Jan 28, 2024, from https://hbr.org/2022/03/dear-manager-youre-holding-too-many-meetings

Layson, M. D., & de Almeida Neto, A. C. (2023). *Development of a Bio-Psycho-Social-Spiritual framework for preventing stress in first responders: Technical Report*. NSW Centre for Work Health & Safety. whs-research/development-of-a-bpss-framework-for-preventing-stress-in-first-responders.pdf

Lehrmann v Network Ten Pty Limited, (2024). (Trial Judgment) [2024] FCA 369.

Litz, B. T., Stein, N., Delaney, E., Lebowitz, L., Nash, W. P., Silva, C., & Maguen, S. (2009). Moral injury and moral repair in war veterans: A preliminary model and intervention strategy. *Clinical Psychology Review*, *29*, 695–706. https://doi.org/10.1016/j.cpr.2009.07.003

Lleó de Nalda, Á., Guillén, M., & Gil Pechuán, I. (2016). The influence of ability, benevolence, and integrity in trust between managers and subordinates: The role of ethical reasoning. *Business Ethics: A European Review*, *25*(4), 556–576. https://doi.org/10.1111/beer.12117

Mandela, N. (2012). *Notes to the future: Words of wisdom*. Simon and Schuster.

Margalit, A. (2017). *On betrayal*. Harvard University Press.

Martinek, N. (2025a, 26 May 2025). The Scapegoating Playbook at work. *Hacking Narcissism*. www.hackingnarcissism.com/p/scapegoatingplaybook

Martinek, N. (2025b, 9 May 2025). So you're the Scapegoat. Now what? *Hacking Narcissism*. www.hackingnarcissism.com/p/scapegoat

Maslach, C., & Leiter, M. P. (2016). Burnout. In G. Fink (Ed.), *Stress: Concepts, cognition, emotion, and behavior* (pp. 351–357). Academic Press. https://doi.org/10.1016/B978-0-12-800951-2.00044-3

Mayer, R. C., & Gavin, M. B. (2005). Trust in management and performance: Who minds the shop while the employees watch the boss? *Academy of Management Journal*, *48*(5), 874–888.

McComas, H. (2023). *Fire and emergency service leadership: Fact or fiction? An investigation into learning and development options in leadership for volunteer firefighters*. Winston Churchill Trust. https://churchilltrust.my.salesforce-sites.com/api/services/apexrest/v1/image/?Id=06998000001prgXAAQ&forceDownload=Yes

McCreary, D. R. (2019). *Veteran and first responder mental ill health and suicide prevention: A scoping review of prevention and early intervention programs used in Canada, Australia, New Zealand, Ireland, and the United Kingdom*. Donald McCreary Scientific Consulting.

Mehnert, J. (2022). *Impact of departmental procedural fairness on officer psychological well-being and engagement in organizational citizenship behaviors* [Doctor of Public Administration, West Chester University].

Mercer, D., Haddon, A., & Loughlin, C. (2018). Leading on the edge: The nature of paramedic leadership at the front line of care. *Health Care Management Review*, *43*(1), 12. https://doi.org/10.1097/HMR.0000000000000125

Michel, J.-B., Yuan Kui, S., Pinker, S., Nowak, M. A., Aiden, E. L., Aiden, A. P., Veres, A., Gray, M. K., Pickett, J. P., Hoiberg, D., Clancy, D., Norvig, P., & Orwant, J. (2011). Quantitative analysis of culture using millions of digitized books. *Science*, *331*(6014), 176–182. https://doi.org/10.1126/science.1199644

Milner, A., Page, A., & Lamontagne, A. D. (2013). Long-term unemployment and suicide: A systematic review and meta-analysis. *PloS one*, *8*(1), e51333. https://doi.org/10.1371/journal.pone.0051333

Nash, S. (2011). Regenerative practice. *Reflective Practice*, *12*(3), 427–438. https://doi.org/10.1080/14623943.2011.571872

Near, J. P., & Miceli, M. P. (1996). Whistle-blowing: Myth and reality. *Journal of Management*, *22*(3), 507–526. https://doi.org/10.1016/S0149-2063(96)90034-3

Noonan, H. W. (2019). *Personal identity* (Third ed.). Routledge. https://doi.org/10.4324/9781315107240

O'neill-George, J. (2022). Betrayal-by-systems: Psychological contract violation and moral injury in Australian Army veterans. Society for Mental Health Research: Breaking down the silos, Hobart, Tasmania.

Office of Parliamentary Counsel. (2017). *Legal Services Directions 2017*. Commonwealth of Australia.

Parker, R. (2008). *The management and operations of the Ambulance service of NSW*. New South Wales Parliament Legislative Council General Purpose Committee No.2. www.parliament.nsw.gov.au/lcdocs/inquiries/1998/100430%20FINAL%20REPORT.pdf

Pyy, I. (2021). Developing political compassion through narrative imagination in human rights education. *Human Rights Education Review*, *4*(3), 24–44. https://doi.org/10.7577/hrer.4482

Ricoeur, P. (1986). Life: A story in search of a narrator (J. N. Kraay & A. J. Scholten, Trans.). In M. C. Doeser & J. N. Kraay (Eds.), *Facts and values: Philosophical reflections from Western and Non-Western perspectives* (pp. 121–132). Springer.

Rousseau, D. M. (1989). Psychological and implied contracts in organizations. *Employee Responsibilities and Rights Journal*, *2*, 121–139. https://doi.org/10.1007/BF01384942

Rousseau, D. M., & Schalk, R. (2000). Introduction. In D. M. Rousseau & R. Schalk (Eds.), *Psychological contracts in employment* (pp. 1–28). Sage Publications Inc.

Royal Commission into Defence and Veteran Suicide. (2024). *Royal Commissioners call on Australians to 'fight for our veterans' ahead of Final Report* https://defenceveteransuicide.royalcommission.gov.au/news-and-media/media-releases/royal-commissioners-call-australians-ahead-final-report

Rule of Law Education Centre. (n.d.). *Principles of the Rule of Law*. Retrieved 2 August 2025 from www.ruleoflaw.org.au/principles/

Safe Work Australia. (2018). *Submission to the Senate inquiry into the high rates of mental health conditions experienced by first responders, emergency service workers and volunteers*.

Safe Work Australia. (2024). *Key Work Health and Safety Statistics Australia: 2024*. Safe Work Australia. https://data.safeworkaustralia.gov.au/sites/default/files/2024-09/Final%20-%20Key%20WHS%20Stats%202024_18SEP.pdf

Schauer, F., & Zeckhauser, R. J. (2007). *Paltering* (Faculty Research Working Papers Series, Issue RWP07-006). https://deliverypdf.ssrn.com/delivery.php?ID=884112031001099106095019112018005024001024032007049053005120125102085119088112085121124025056115114005124116024101100096106009043043051022126029116005099001031022060075092079090090010001096119122117023007127095117087013025010029029103098112078119&EXT=pdf&INDEX=TRUE

Shay, J. (1994). *Achilles in Vietnam: Combat trauma and the undoing of character*. Scribner.

Shay, J. (2002). *Odysseus in America: Combat trauma and the trials of homecoming*. Scribner.

Shay, J. (2014). Moral injury. *Psychoanalytic Psychology*, *31*(2), 182–191. https://doi.org/10.1037/a0036090

Smidt, A. M., Adams-Clark, A. A., & Freyd, J. J. (2023). Institutional courage buffers against institutional betrayal, protects employee health, and fosters organizational commitment following workplace sexual harassment. *PloS one*, *18*(1), e0278830. https://doi.org/10.1371/journal.pone.0278830

Smith, C. (2003). *Moral, believing animals: Human personhood and culture*. Oxford University Press.

Smith, C. P., & Freyd, J. J. (2014). Institutional betrayal. *The American Psychologist*, *69*(6), 575–587. https://doi.org/10.1037/a0037564

Smith, G. (2018). *Paper safe: The triumph of bureaucracy in safety management*. Weyland Legal.

Smith-MacDonald, L., Lentz, L., Malloy, D., Brémault-Phillips, S., & Carleton, R. N. (2021). Meat in a seat: A grounded theory study exploring moral injury in Canadian public safety communicators, firefighters, and paramedics. *International Journal of Environmental Research and Public Health*, *18*(22), 12145. https://doi.org/10.3390/ijerph182212145

Snow, J. (1849). On the mode of communication of cholera. London: John Churchill.

Sullivan, M. J. L., Adams, H., Horan, S., Maher, D., Boland, D., & Gross, R. (2008). The role of perceived injustice in the experience of chronic pain and disability: Scale development and validation. *Journal of Occupational Rehabilitation*, *18*(3), 249–261. https://doi.org/10.1007/s10926-008-9140-5

Sydney Diocese Standing Committee. (2024). Use of non-disclosure agreements in the Diocese: 48/23 Non-Disclosure Agreements. In *Second Session of 53rd Synod: Proceedings for 2024* Sydney Diocesan Secretariat. https://orgs.sydneyanglicans.net/s/sfsites/c/sfc/servlet.shepherd/document/download/069O100000KpR5tIAF

Taylor, C. (1992). *The ethics of authenticity*. Harvard University Press.

Tedeschi, R. (2004). Posttraumatic growth: Conceptual foundations and empirical evidence. *Psychological Inquiry*, *15*(1), 1–18. www.jstor.org/stable/20447194

Tedeschi, R. G., Cann, A., Taku, K., Senol-Durak, E., & Calhoun, L. G. (2017). The posttraumatic growth inventory: A revision integrating existential and spiritual change: Posttraumatic growth inventory and spiritual change. *Journal of Traumatic Stress*, *30*(1), 11–18. https://doi.org/10.1002/jts.22155

Tedeschi, R. G. R. G. (1996). The posttraumatic growth inventory: Measuring the positive legacy of trauma. *Journal of Traumatic Stress*, *9*(3), 455–471. https://doi.org/10.1002/jts.2490090305

The Honourable Spigelman AC, J. J. (2006). *Foreword*. www.judcom.nsw.gov.au/publications/benchbks/equality/foreword.html

Thorndike, E. L. (1920). A constant error in psychological ratings. *Journal of Applied Psychology*, *4*(1), 25–29. https://doi.org/10.1037/h0071663

Victorian Equal Opportunity & Human Rights Commission. (2022). *Independent review into workplace equality in Ambulance Victoria: Final report (Volume 2)* (0648088227). www.humanrights.vic.gov.au/static/5631441a7f206c4abe904a040db49254/Resource-Ambulance_Victoria_Independent_Review_Workplace_Equality-Volume_2-Final-Report_v2.pdf

Volf, M. (2019). *Exclusion and embrace: A theological exploration of identity, otherness, and reconciliation* (2nd ed.). Abingdon Press.

Voss, C., & Raz, T. (2016). *Never split the difference: Negotiating as if your life depended on it*. Random House.

Wadham, B., & Connor, J. (2022). *Research into defence abuse 2018-2022 (Submission to Royal Commission into Defence and Veteran Suicide)*. Flinders University | Orama Institute. https://static1.squarespace.com/static/5829ccde2e69cf19589499ac/t/627a2087e902913a971a342a/1652170892651/Submission+to+Royal+Commission+2021+%281%29.pdf

Whyte, M. (2008). The uses and abuses of Nietzsche in the Third Reich: Alfred Baeumler's 'Heroic Realism'. *Journal of Contemporary History*, *43*(2), 171–194. www.jstor.org/stable/30036502

Woodhead, L. (2022). Truth and deceit in institutions. *Studies in Christian Ethics*, *35*(1), 87–103. https://doi.org/10.1177/09539468211051162

World Health Organization. (2022). *WHO guidelines on mental health at work*. World Health Organization. https://iris.who.int/server/api/core/bitstreams/6152a556-6893-4c4e-9ed8-094478bb25eb/content

Index

O

P

Q

R

S

For Product Safety Concerns and Information please contact our EU
representative GPSR@taylorandfrancis.com
Taylor & Francis Verlag GmbH, Kaufingerstraße 24, 80331 München, Germany

www.ingramcontent.com/pod-product-compliance
Ingram Content Group UK Ltd.
Pitfield, Milton Keynes, MK11 3LW, UK
UKHW022125190926
475702UK00013B/1646